TIES
TEACHER INSTITUTE FOR
EVOLUTIONARY SCIENCE

ON TEACHING EVOLUTION

Edited by

Bertha Vázquez
www.tieseducation.org

Foreword by **Richard Dawkins**

KEYSTONE
CANYON PRESS

Keystone Canyon Press
2341 Crestone Drive
Reno, NV 89523
www.keystonecanyon.com

Publisher Alrica Goldstein
Copyeditor Paul Szydelko
Cover Design Alissa Booth
Indexer Christine Hoskin

Library of Congress Control Number: 2021938581

ISBN 978-1-953055-25-5
EPUB ISBN 978-1-953055-26-2

Manufactured in the United States of America

For every single science teacher who has presented evolution

in their classroom for what it is—the awe-inspiring, unifying theme

of the life sciences.

We invite readers to visit tieseducation.org/book to find all of the videos and resources mentioned throughout this book and to support your local bookstore by purchasing suggested books at bookshop.org/lists/on-teaching-evolution.

Contents

Foreword by Richard Dawkins...vii

Introduction 1

1   For Patricia Soto by Bertha Vázquez 7

2   The Accidental Evolution Teacher by Nikki Chambers......................15

3   The Cherokee Creation Myth by Amanda Clapp............................ 23

4   Showing Evidence of Evolution by Kenny Coogan 29

5   Teaching Darwin's Theories by Robert A. Cooper........................... 33

6   An Ode to the Unbroken Thread by Chance Duncan......................41

7   We Are All Cousins by Reginald Finley, Sr. 47

8   Cultural Border Crossing by Katie Green................................... 55

9   The Evolution of an Evolution Advocate—A Lifelong Journey
by John S. Mead 61

10 Chipmunks, Eternal Damnation, and Other Hazards of Biology
by David Mowry.. 73

11 The Nature of Science, Evolution, and Storytelling by Blake
Touchet 81

12 Evolution and a More Just Society by David Upegui 93

13 A Compassionate Worldview by Patti Howell 103

Conclusion: Looking to the Future.. 111

Notes on Activities..112

Bibliography 113

Index..117

Author Biographies..124

Acknowledgments 127

If you want to build a ship, don't drum up people to collect wood and don't assign them tasks and work, but rather teach them to long for the endless immensity of the sea.

Antoine de Saint-Exupéry

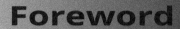

Foreword

Richard Dawkins, PhD

"If you can read this, thank a teacher." So runs a favorite T-shirt slogan. This book inspires me to coin a variant: "If you understand why you exist and rejoice in that understanding, thank a science teacher." More specifically, thank a teacher of evolution. It is a remarkable fact that before Darwin and Wallace burst on the scene, humanity had no sensible account of why we exist—what life is all about. Before Darwin's foundational work, people just supinely and incuriously accepted the fact of their own existence. Not to mention the existence of the dazzling riot of life all around them, from green plants to elephants, from ants to the Great Barrier Reef.

Darwin changed all that. Evolution is perhaps the most thrillingly eye-opening subject any student is ever called upon to learn. Yet, as several of these chapters testify, many students have been brought up by their parents to mistrust the subject, uniquely among all topics in science. This has the effect of compelling teachers to exercise extreme sensitivity and cautious diplomacy. Where teachers of other subjects walk with confident steps into the classroom, teachers of evolution walk on eggshells. In their different ways, the authors of this book have risen to the challenge. They have refrained from evasions such as the contemptible resort of a colleague who "avoided the topic altogether, because it wasn't worth the hassle." You might as well avoid verbs when teaching French!

Mention of verbs calls to mind a lovely notion, which I met for the first time in the pages of this book. "Science as a verb." Science as something you do, rather than as a list of facts to be memorized. Many of these chapters tell of highly imaginative lessons designed to inspire the students with hands-on experience, calling upon them to think creatively and ingeniously. Enthusiasm and a sparkling love of their subject are two of the greatest gifts a teacher can bring to the classroom. These qualities are much in evidence in this book, summed up in a memorable phrase from one of the authors. A teacher who is "happy as a clam in an intertidal zone" will excite students to the same happy and productive state.

Returning to the peculiar problem faced by biology teachers, one of our authors correctly said, "I could not possibly imagine how a teacher could educate a student about biology without the unifying, underlying theme of evolution. It ties concepts together in a way that is both simple and wildly complicated. Without this theme, biology would just be a list of random disconnected facts for students to memorize." It has always struck me as odd that textbooks of biology so often relegate evolution to the final chapter. It should, of course, be chapter 1, for none of the other chapters will make any sense without it.

I reluctantly have to confess that I don't think I'd be very good at coping with the active, religiously motivated pushback that these teachers have to face. "Evolution sounds like an interesting theory, but unfortunately, I don't believe in it." This farcical statement from a student, quoted here somewhere, had me reeling. How would I respond to it? You might as well not believe in gravity—in which case my inclination would be to point to an upstairs window and invite the student to jump. Evolution is not only interesting, it's true! As you'd see, if only you'd open your eyes and look at the evidence. Open your mind and think about the evidence; for evidence is, or should be, the only basis for believing in anything. No, I'm afraid I'd be a failure as a teacher of students like that, which makes me respect the long suffering patience of these authors all the more.

Several chapters allude to the great "only a theory" problem. For years I went along with the standard catechism: science uses "theory" in a different sense from everyday speech where it means "tentative hypothesis." While that is accurate, I'm afraid we are failing to put it across, and the reason is that it actively begs to be misunderstood. "Only a theory" remains undented as the most powerful weapon in the creationist's mendacious armory. I've now switched tactics. Evolution is a fact. It's a fact in the same sense as it's a fact that the planets orbit the sun, and it's every bit as secure.

To be sure, philosophers of science will protest that any fact is only a hypothesis that has so far not been disproved. I like Stephen Gould's retort: "In science, 'fact' can only mean 'confirmed to such a degree that it would be perverse to withhold provisional assent.' I suppose that apples might start to rise tomorrow, but the possibility does not merit equal time in physics classrooms." "Provisional" is an understatement. To deny the fact of evolution in the face of all the evidence (especially molecular evidence) that has been added to Darwin's own massive compilation, would be as perverse as to deny that viruses and bacteria cause disease. We know for a fact that we are cousins of kangaroos, kinkajous, and kookaburras. This "theory" is as likely to be disproved as the theory that apples will rise tomorrow.

Evolution is a fact, but it is still appropriate to use "theory" for natural selection as the dominant driving force in adaptive evolution. There are other driving forces, and they are important: genetic drift, for example. But natural selection remains the only theory ever proposed that is in principle capable of generating functional adaptation. In Darwin's time it was appropriate for him to refer to evolution as a theory in the colloquial sense of hypothesis. You can still call it a theory if you insist, but you'll be widely and mischievously misunderstood. The evidence has now built up beyond the point where to call it anything but a fact would be perverse. Let's all stop calling it a theory and call it what it is. A fact.

I can't name all the individual teachers who have contributed to this splendid book, and I have therefore consciously refrained from naming any of them. But I have to make one exception. Bertha Vázquez is the driving inspiration of the Teacher Institute for Evolutionary Science and of this book. She is a teacher in a million. You have only to see the light of enthusiasm in her eyes, hear the dedication in her voice, to know she would captivate students. But more than that, she works hard behind the scenes to create model lessons and resources not just for her own classes in Florida but for the whole TIES community that she has inspired in every state of the country.

Introduction

Bertha Vázquez, MS in Science Education
Miami, FL

Ever since the famous (infamous) John Scopes trial in 1925, teachers have been at the forefront of the fight to have evolution taught properly in America's schools. While some teachers have been literally named in famous court cases, *Epperson v. Arkansas,* for example, most teachers are silent fighters. We never hear from them. And make no mistake, it is a daily challenge to teach the children of this great country one of the cornerstones of scientific thought. While politicians and school board members make it on the news, swinging the evolution education pendulum this way and that, it is our nation's science teachers who spend their days with children, putting in the hours and making a difference. These science teachers can inspire America's future generations and help them understand the importance of evolutionary biology in everything from agriculture and medicine to conservation and human behavior. These foot soldiers introduce students to the "grandeur of this view of life." To see the world through the lens of evolutionary biology is to see the awesome connectedness of all nature. This book provides some of these worthy educators a voice. Each chapter is written by a different teacher, a different voice. Our backgrounds and teaching positions are diverse, but we all share something in common. We care. We understand the impact of what we do every day. All the teachers who have contributed to this book are members of the Teacher Institute for Evolutionary Science (TIES). Our collective passion for the topic of evolution is what brought us together. I thank them all for collaborating on this worthy project.

History of Evolution Education in the US

The history of evolution education in this country goes back at least a century. Before we hear from the contemporary voices in this book, some background is in order.

Evolution vs. Creationism (2004) by Eugenie C. Scott and the fine compilation edited by Andrew J. Petto and Laura R. Godfrey titled *Scientists Confront Creationism: Intelligent Design and Beyond* (2008) are excellent books, thoroughly researched sources detailing the embattled history of evolution education in the United States. Since these in-depth studies on the history of evolution education are readily obtainable, we will only briefly discuss them in this book.

The history of evolution education can be delineated with the following progression during the last ninety years: **Phase 1—Banning Evolution Education:** The Scopes trial in Tennessee in 1925 led to at least two decades when evolution was hardly mentioned in science textbooks at all. As a matter of fact, evolution curriculum was found in our nation's high school classrooms more often before the Scopes trial than after it. In the 1950s, the space race with the Soviets sounded the alarm for the importance of science education, including evolution. It wasn't until 1968, however, in the case of *Epperson v. Arkansas,* that it became unlawful to ban the teaching of evolution.

Phase 2—Equal Time Laws: The battle against evolution education took on a different theme in the late 1960s and throughout the 1970s. If evolution education could not be banned, its opponents argued for "equal time" for alternative religious explanations of life on Earth. In 1981, Arkansas Act 590 was the first piece of this so-called "equal time" legislation to pass into law. Twenty-seven other states had tried to pass similar legislation before Arkansas without success. In *McLean v. Arkansas Board of Education,* the Arkansas law was challenged and the law was declared unconstitutional. A similar equal time law was passed in Louisiana in 1982. After years of litigation, delays, and appeals, the US Supreme Court finally ruled against equal time for particular religious viewpoints being taught alongside biological evolution in classrooms. (*Edwards v. Aguillard* 1987).

Phase 3—Repackaging Creationism as Science: The next phase of the struggle was the fight against the promotion of intelligent design. Intelligent design is creationism offered up as an alternative scientific theory. When speaking or writing about intelligent design, its defenders avoided referring to religion or creationism altogether, just suggesting that evolution was not the only viable scientific explanation for how life has changed over time on our planet was enough. *Kitzmiller v. Dover Area School District* (2005) in Pennsylvania ended this round of the pseudo-controversy with the judgment of Judge John E. Jones, "The citizens of the Dover area were poorly served by the members of the Board who voted for the ID [intelligent design] Policy. It is ironic that several of these individuals, who so staunchly and proudly touted their religious convictions in public, would time and again lie to cover their tracks and disguise the real purpose behind the ID Policy." (2005)

Phase 4—Academic Freedom: We now find ourselves in the current stage of the battle, the promotion of "academic freedom" bills. Essentially, these bills allow a teacher to insert creationism into classroom discourse by presenting all the so-called scientific views on a topic. It also allows teachers to discuss the strengths and weaknesses of scientific theories. Of course, teachers do not

seem to be exercising their "academic freedom" to debate the merits of whether doctors should wash their hands before surgery, a practice strongly encouraged by the overwhelming evidence behind the germ theory. Nor are they allowing their students to analyze the fine points of the ongoing natural selection debate between Richard Dawkins and E. O. Wilson. Not surprisingly, these bills seem to only address such "controversial" topics such as climate change and the age of Earth.

The first academic freedom bill was introduced in Alabama in 2004. From 2004 to 2015, at least eighty academic freedom bills have been filed in eighteen states. Only three have passed, in Tennessee, Mississippi, and Louisiana. A current, detailed list of these bills can be found under "Academic Freedom" Legislation on the National Center for Science Education webpage. (Chronology of "Academic Freedom" Bills 2013).

Despite this constant struggle against anti-science forces, current trends in evolution acceptance among young people in the United States are encouraging. For example, the Pew Research Center found that younger adults are more likely than older generations to believe that living things have evolved over time. (Pew Forum 2013)

The cause for this change in evolution acceptance may be an overall loss in religiosity, but perhaps it also is because this younger group of adults has received a better evolution education. "Over the past decade, the concerted efforts of various academic and scientific organizations have led to greater emphasis in textbooks and curricula on the central place of evolution in understanding life" (Carroll 2014).

So, yes, education makes a difference. Yes, the educators who have contributed to this book are on the front lines of this effort. Understanding evolution will be vital as scientific research continues to advance. Let's hear the very personal stories of the teachers who are educating the next generation of Americans.

History of the Teacher Institute for Evolutionary Science

The purpose of the Teacher Institute for Evolutionary Science is to familiarize interested middle school science teachers with the concepts of natural selection, common ancestry, and diversity for them to confidently cover the

Richard Dawkins speaking to teachers in Miami, Florida in 2014.

topics in their classrooms and fulfill their curriculum requirements. TIES introduces middle school teachers to the most important points of evolution and natural selection with a focus on the amazing advances of genetics. The success of TIES depends upon providing resources that teachers can begin to use immediately. Participating teachers or student teachers leave our workshops with presentation slides, labs, guided reading assignments, an exam, and a valuable resource list for their lesson plans. Our webpage is a one-stop shop for evolution education, and we constantly add new resources on our Facebook page as well.

Let me offer some history on the Teacher Institute for Evolutionary Science and why I decided to focus on middle school science teachers. A middle school science teacher is our system's jack of all trades. It is virtually impossible to become an expert in all our content areas, at least not initially. I have taught everything from meteorology to the laws of motion. I have often stayed a chapter ahead of my students as I learned the difference between an occluded front and a stationary front in the unit on meteorology. And, over the years, it has repeatedly dawned on me that my greatest resource for learning new material and developing effective lesson plans has been my fellow middle school science teachers. We are a talented bunch. My third and fourth years in the classroom were a magical time for me. I team-taught more than sixty sixth graders with one of my school district's shining stars, Mrs. Patricia Soto. Mrs. Soto could captivate a room full of eleven-year olds with any subject matter; her focus was hands-on learning and science inquiry. To this day, I find myself using her ideas and strategies countless times a school year. Years later, I once again found myself team-teaching with an exceptional educator, Mary Martinez. She loved geology and had an extensive rock and mineral collection at her disposal. I learned the wonders of asterism and chatoyancy right alongside the children. If the students only knew that I had no idea what kind of mineral they were showing me, "Hey, Ms. V, is this feldspar?"

"Hmmm," I would answer. "What do you think it is?" As the students studied the mineral, I would walk over to Mary and ask, "Mary, what the heck is this?" When the students were not looking, I would lick the white minerals, knowing I could identify it as halite if it tasted salty.

Besides recognizing the invaluable resources my colleagues provide, I realized we teach best what we know and love best. Our knowledge of a subject leads to our own enthusiasm for it, and this makes a significant difference in our students' learning process. Passion is contagious. Believe me, you want your children learning geology from Ms. Martinez, not me.

Like the countless teachers who have kindly opened their file cabinets and generously offered me lesson plans and lab activities, I wanted to provide something meaningful for my fellow science teachers. Science understanding is constantly expanding. It is very difficult for science teachers to keep up with all the latest research across all the subject areas they teach. I became more interested in providing teachers with professional development opportunities in evolution education specifically after an exciting afternoon with the person many consider today's living representative of Charles Darwin, Richard Dawkins. Professor Dawkins had been invited to be a professor at the University of Miami for one week in the spring of 2013. I was very fortunate to be invited to join the students and faculty of the UM Biology Department for all his small-group lectures.

Now, I must stop and explain that Dawkins has been the greatest intellectual influence in my life. I read his seminal work, *The Selfish Gene*, back when I was in college, and it completely shifted my paradigm. His logic and clarity of thought were very appealing and insightful. I read all his books. And while *The Selfish Gene* was instrumental in making me choose biology as a college major, it was *The Ancestor's Tale* that really captivated me. The thought experiment proposed in

this book, of going back in time and tracing our ancestry while meeting our fellow pilgrims along the way, all making our way toward the origin of life itself, was wondrous to me.

Needless to say, it was a thrill to find myself sitting next to him at a small lunch table following one of his lectures. Dawkins was listening intently as one of the biology professors was explaining how evolution could no longer be taught at his son's school. The owner of this local private school had banned teachers from teaching evolution because one parent had complained. We discussed how middle school science teachers in America may not necessarily be well-versed in evolutionary biology and, with this kind of pressure from parents and community leaders, they simply skip over it. Teachers are by nature risk-averse. Would they teach evolution more effectively if they had confidence in the subject and good resources at their disposal?

I went back to my classroom with a mission. Here's a topic that I loved and I had files full of good lessons. Therefore, in 2013, I offered my fellow science teachers a series of workshops on evolution. The highlight of the sessions was a guided discussion of the wonderful book, *Your Inner Fish,* by Neil Shubin.

Dawkins was back in Miami in November 2014 and I shared my experiences with him. He intuitively understood the importance of giving the teachers of this impressionable age group the proper tools to teach evolution. In what was truly a testament to his commitment to science education, he offered to come to my middle school on December 11, 2014, and speak with middle school teachers from all over Miami-Dade County on the Florida State Science Standards on Evolution and Natural Selection. In a two-hour interview, he and I addressed the fundamentals of evolutionary science. Kudos to Miami-Dade County Public Schools for opening the event up to all the school district's science teachers.

Based on the responses from the teachers who attended, the afternoon was a great success. The responses alerted us to the need for middle school science teachers to properly understand evolutionary biology.

This revelation was the cornerstone of the creation of the Teacher Institute for Evolutionary Science (TIES). I was offered the position of director of this new, exciting project. The Richard Dawkins Foundation for Reason

& Science could provide me with the resources to make professional development in evolutionary biology for middle school science teachers an ongoing endeavour.

I decided right away that while the main goal of TIES would be to promote effective evolution education at the middle school level, several other equally significant goals would be addressed. First, our rationale would be to promote the idea that evolutionary biology is awe-inspiring. Teaching evolution enables children to make sense of the world around them. It provides them with an understanding of how all life on Earth is related. Not only is this knowledge exciting, it is the key to many current conservation efforts, agricultural practices, and medical breakthroughs. Understanding evolution is essential if the United States is to continue to be a global leader. I think this spirit is reflected in the chapters which follow.

This was not the only goal I had in mind. TIES would promote teacher leadership. Sadly, the most important decisions in education in the United States often are made by those who aren't teachers. Classroom teachers would present TIES workshops. It takes a teacher to know what another teacher is experiencing daily. I would work with these teacher presenters to compile a list of classroom resources and online activities.

TIES workshops and resources would all be free. TIES materials include presentation slides with active learning ideas, hands-on activities, guided readings, and informative videos. Extremely valuable online resources and recommended readings would also be included. And finally, TIES workshops would not present a single lab activity or hands-on lesson but provide teachers with an entire unit of instruction, from the ice breaker activity to the exam.

TIES would also provide valuable resources of professional support. For example, the National Science Teachers Association (NSTA) has issued an excellent position statement on evolution education. Other sources, such as the National Science Foundation and the comprehensive Understanding Evolution website of the University of California at Berkeley, can become essential when unhappy parents confront science teachers. Teachers can demonstrate that they are not the ones responsible for setting the curriculum for the class and the parents will have to go elsewhere to complain. Having access to these resources takes the teacher "off the hook," so to speak.

Bertha Vázquez with a Silver Fox from the famous domestication experiment in Siberia.

TIES curriculum would highlight modern-day examples of evolution. Sadly, many students are automatically turned off by Darwin's name and anti-evolutionists have deliberately and falsely tried to discredit iconic examples of evolution (Gishlick 2003). A powerful example of the influence TIES can have in a classroom with this goal in mind occurred in April of 2016. I teamed up with a biology professor, Eric von Wettberg, formerly of the Florida International University. One of the participating teachers introduced herself by telling us that she didn't really believe in evolution. She explained that she tells her students that they must study and get a good grade on the test so she can move on. I presented the standard TIES content in the morning, and von Wettberg discussed his research in the afternoon. He explained that 20 percent of the world's population relies on the chick pea for its primary source of protein

Teachers at the inaugural TIES workshop.

and that the global yield of the chickpea crop is declining because of climate change. His lab is attempting to cross the agricultural strains of the chickpea with the much more robust wild strain still found today in southern Turkey and northern Iraq. By introducing genetic variation into the agricultural strain, he is making it a hardier, more resilient crop. In other words, he is using the principles of natural selection to ensure that millions of people continue to have access to an important food source. Our disbelieving teacher left the workshop with a totally different perspective. We can be optimistic that her students will be receiving a very different view of evolution in her classes.

Many TIES workshops, like the one described above, have been the result of helpful collaborations with other community partners, such as zoos, museums, and universities. The first TIES workshop took place thanks to one of these important collaborations with the Phillip and Patricia Frost Museum of Science in Miami. On April 3, 2015, thirty middle school science teachers participated in a daylong workshop that included several guest speakers from the staff of museum scientists. The museum scientists shared research on phylogeny, fossilized amber, the fossil preparation of Xiphactinus (a 475-million-year-old fish), and provided a sneak peek of future museum exhibits. And, as a most special treat, the participants played with a pet silver fox. These foxes are the result of fifty years of artificial selection experiments in Siberia. After allowing only the tamest foxes to breed every generation for fifty years, Russian scientists have created a breed of fox with markedly decreased stress hormone levels. This has created a gentle group of foxes who resemble dogs in both physical and behavioral characteristics.

Since this first workshop more than six years ago, TIES has expanded into all fifty US states. More than ninety classroom teachers have presented more than three hundred workshops varying from one hour to three-day events. Our free resources have been downloaded more than three thousand times. We have also launched the free TIES webinar series, in which biologists and science authors are featured monthly.

Within every experienced classroom teacher is a wealth of pedagogical and content knowledge just waiting to be tapped. We are our own best resources. May you find some excellent ones within these pages.

1

For Patricia Soto

Bertha Vázquez, MS in Science Education
Miami, FL

Albi, France
September 1989

I nervously looked around the classroom as the high school students filed in. I was a twenty-one-year-old college graduate wondering what in the world I had gotten myself into. The idea of spending a year teaching English in France after graduation seemed like a great idea when my French professor mentioned it a year earlier. After all, I had already done a summer semester at La Sorbonne in Paris and truly enjoyed French culture. "Do it now!" everyone said. "You do not have a steady job yet, no boyfriend, no responsibilities. Enjoy a year in France." Anybody could teach, right? Everyone sits in a classroom for more than a decade, how hard could it be? Thankfully, my classes were small, and my students were willing. I had no set curriculum to cover and even taught some of the classes at the nearby café sometimes. My students and I sat around cups of coffee and discussed current events, music, sports, history. From the very first day, I knew it was in me . . . to teach. I loved the exchanges with the students, the feeling of guiding them along their academic journeys, of making them appreciate American culture and the English language. I was hooked.

When I returned to Miami, Florida, in August 1990, I knew I wanted to be a teacher. I majored in biology in college. Teaching science was the natural choice. Like many people, I imagined myself working in a high school and only teaching what I knew best, biology. But, alas, middle school is where many new teachers end up, and it's where I began my career as a public school science teacher. In many places, math and science teachers are in such high demand that people are hired on the spot, without the required education certification. This was my case. I was given two years by the school district to earn my education credits and pass the state certification exams. I was incredibly lucky to end up teaching at a school with a master teacher named Patricia Soto. She generously shared her knowledge, her passion, and her resources with me. I can still hear her voice in my head when I'm teaching. Eight years later, I achieved national board certification and went back to school to earn my master's degree in science education.

Being a science teacher is the greatest job on earth. Science can be a truly wondrous gift to share with young people. Although the day-to-day interaction can be exhausting, often making me feel like I am being pecked to death by ducks, I do not regret my decision thirty years ago to become a science teacher. I love working with young people and introducing them to humanity's best way of finding answers. On the first day of school, I have often asked students to draw a scientist. It never fails, I get drawing after drawing of an old man with crazy hair and a lab coat, standing passively with a table of colorful chemicals before him. Is this what students think scientists look like? I've gotten this drawing from students of all ages, from sixth graders to college students. Honestly, it depresses me every year.

It's our job as science teachers to erase this image from the minds of our young people and to replace it with pictures of themselves. To replace it with somebody looking at the world with a mix of scepticism and wonder. At the end of the school year, I often try this activity again. The pictures on the last day of school are quite different from those just ten months earlier. You find scuba divers, food chemists, car engineers, and volcanologists. The people in the pictures no longer stand passively; they are observing, collecting, and collaborating. Many of the pictures show young people studying plants and animals, comparing their DNA and building family trees. Yes, evolutionary biology is a very appealing topic when presented with the depth and breadth it deserves. Evolution truly provides the logical framework for the biological sciences; it makes everything from biochemistry to ecology fit together perfectly. Sadly, teachers are faced with many challenges when they try to teach evolution properly.

Challenges of Teaching Evolution in the US

One of the most important things teachers can do is build rapport with their students. Because I have my students for at least two consecutive years, we develop a relationship built on trust and mutual respect over time. In my classroom, learning takes place in a welcoming environment. This firm but friendly atmosphere is only compromised during one unit of study, the unit on evolution.

Richard Dawkins said it best in his book, *The Greatest Show on Earth: The Evidence for Evolution*. Imagine being a professor of Roman history who must constantly, year after year, defend the very existence of the Roman Empire. Despite the overwhelming evidence coming from various sources, architecture, art, literature, etc., your students are not only disbelieving, they can be downright disrespectful. This is what teaching evolution feels like for many science teachers.

Like most teachers, I have at least a handful of students every year who are anxious about learning evolution. For example, many students who are raised in faiths which totally accept evolution ask whether they are even "allowed" to learn it in my class. I've had students refuse to do any assignments related to evolution or who will sit with their backs to me for the entire evolution unit. This creates anxiety, not only in the students who are

told at home that evolution is false but in the other children in the room who do not understand what all the fuss is about. Other students who completely accept and understand what I'm teaching are told by their families or pastors that they should not believe a word I say. Four years ago, an eighth grader came to me in tears after she defended evolution to her pastor who proceeded to call me a "disgusting human being."

Three years ago, a parent confronted me during my school's annual open house. She asked me if I believed in God. She told me that I was causing tremendous anxiety in her son because I began the school year with a series of lessons on the biology of skin color. Luckily for her son, she claimed, their pastor set him straight on evolution. She also stated that evolution was a religion and that there was no evidence for it. The other parents sat motionless, disbelieving what was happening. I remained calm throughout the exchange and responded that evolution was not a religion and that there were mountains of evidence for it. I politely asked her if we could continue the conversation at a more appropriate time. Welcome to the world of the American science teacher.

The figures nationwide are disconcerting. When asked if they felt pressure to teach creationism in their classrooms, 31 percent of high school biology teachers reported that they did. (Berkman and Plutzer, 2010) They reported that this pressure comes primarily from students and parents. These pressures can be very stressful for teachers, who often teach evolution at the end of the year as a discrete topic, instead of as the unifying theme of biology that it is. Many teachers will begin their unit on evolution with the words, "You do not have to believe what I'm about to teach you. You just have to understand it and try your best on the test." One school district in Georgia even went so far as to create an opening activity with this message:

> "This unit typically comes with some backlash from students and parents. It may help at the beginning of the unit to give students an analogy . . . Ask students to turn to a partner and come up with a definition of a zombie . . . Ask how many students believe zombies are real . . . Next, do the same for alien. Explain to the students that the class is going to be learning about

BOOK RECOMMENDATIONS

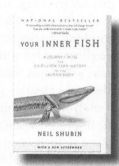

It's impossible for me to single out a single book. There are so many books out there, by so many brilliant authors. For example, *Your Inner Fish: A Journey into the 3.5-Billion-Year History of the Human Body* by Neil Shubin does an excellent job of describing the modern synthesis of evolution, genetics, and embryonic development. You can guarantee a collective jaw drop in your class by telling students that scientists can take tissue from a chicken embryo where a wing is about to develop and place it in just the right spot in a developing mouse embryo. Does the mouse grow a normal front leg or a wing? A normal mouse leg. Why can the mouse continue its normal development with the tissue of a completely different species? Shubin uses example after example to illustrate our common ancestry with other living things. And if the book is too complicated for students, teachers can show parts of the three-hour television series instead.

Another great book that really digs into the idea that nature is far from perfect but rather full of evolutionary dead ends, incomplete projects, and downright crazy mistakes is *Human Errors: A Panorama of Our Glitches, from Pointless Bones to Broken Genes* by Nathan H. Lents. Teachers can download a lesson I wrote based on Lents's book from the TIES website.

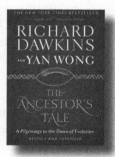

If I were to choose a book based solely on my own personal preferences, it would be *The Ancestor's Tale: A Pilgrimage to the Dawn of Evolution* by Richard Dawkins. In this beautifully written opus, modern living things are pilgrims heading back in time to the origin of life. Along the way, they meet their fellow pilgrims and listen to each other's tales. I've read my own copy so many times it is dog-eared and battered, with notes on practically every page. A great accompaniment for this book is the website, OneZoom, an ambitious project to re-create this family tree of life. Students can tool around the tree of life, finding where all of life's species fall in terms of relatedness and common ancestry.

evolution. We are not telling you that you have to believe in evolution. However, we are asking that you learn enough about evolution to be able to explain the argument or definition to someone else similar to what you did for aliens and zombies. You can learn information about a topic or concept to have a conversation and answer questions without having to believe it." (Troup County schools, Georgia)

I know this sounds absolutely unbelievable to somebody outside of education, but I understand what this school district is trying to achieve. This ice breaker will most certainly help their science teachers avoid confrontation with anyone who does not "believe" in evolution. It's trying to help teachers avoid phone calls and visits from angry parents. Nonetheless, this school district is dispensing unacceptable advice to its science teachers. Both the Center for Inquiry and the National Center for Science Education attempted to contact this school district about this activity, but neither received a response.

Teachers are, by nature, nonconfrontational. We prefer to avoid issues that will involve angry parents and students. This avoidance is even worse in middle school, where it's possible a teacher may not have majored in the subject they are teaching. You see, middle school science teachers have to teach it all, from meteorology and space science to physical science. This means it's possible to have an excellent teacher with no life science background, for example, who must teach evolution. Without a firm grasp of the subject matter, teachers are even less likely to take on a controversial subject head-on.

While improved science education is only one variable involved in increasing evolution acceptance in the United States, it is arguably a very important piece of the puzzle. After looking into the most effective ways to improve evolution education, the National Center for Science Education concluded that "requiring all teachers to complete a course in evolutionary biology would have a substantial impact on the emphasis on evolution and its centrality in high school biology courses. In the long run, the impact of such a change could have a more far reaching effect than the victories in courts and in state governments." (NCSE 2008) Biologist Sean B. Carroll notes that the acceptance of evolution is up to 68 percent among adult Americans under age thirty (it's 60 percent overall). He attributes the increased acceptance among young people to the fact that "over the past decade, the concerted efforts of various academic and scientific organizations have led to greater emphasis in textbooks and curricula on the central place of evolution in understanding life." (Carroll 2014)

As a science teacher myself, I have seen firsthand how a good teacher can improve a student's understanding of a science topic. Many school principals will tell you that they prefer hiring a good teacher with no science background rather than a scientist with no teaching experience. A good teacher can deliver, once the content knowledge and resources are provided to them. The key is giving them the necessary content knowledge. This excerpt from Caitlin Schrein's interview with Stephanie Keep of the National Center for Science Education absolutely sums up this opinion:

> Teachers need to feel confident teaching the subject, as well. This confidence can come from education and training, but also from access to experts and high quality teaching resources. In my opinion, teachers who are really at the top of their game can teach just about anything as long as they have the resources to do so. There are many enthusiastic and knowledgeable science teachers who don't know specific areas of science, like evolutionary theory, through and through, but who are perfectly capable of teaching those subjects effectively when properly supported. (NCSE 2016)

I believe Schrein hit the nail on the head with this quote and recent studies prove it. Two studies published in the last year by NCSE show that evolution education is improving in both middle and high school science classes around the country. Teachers are spending more time on evolution curriculum and less teachers are including non-science content in their lessons, such as intelligent design. (Branch, Reid, and Plutzer 2020 and 2021) The key to these improvements is high-quality professional development for all science teachers. No doubt that many factors such as those mentioned by Carroll are contributing to this improvement. I would like to think that the TIES project with its amazing reach into all fifty US states has had something to do with that, no matter how small!

Introducing Evolution

Many students walk into class with preconceived notions about evolution. Even students who are ready to accept evolution at face value know that some people, somewhere, take issue with the topic. They often want to know why. Before I answer any questions, my goal is to introduce evolution (natural selection, actually) in the simplest terms possible. That is why I created the bell-ringer, "Those Pesky Fleas." To use this bell-ringer, students must have a basic understanding of genetics. The pacing or curriculum guides for many school districts cover evolution first. If that is the case in your area, I propose teaching a lesson on genetics before starting your evolution unit. The TIES classroom presentation includes several slides on basic genetics and can be downloaded for free.

On the first day of the evolution unit, I show the students a box of flea medicine. I proceed to tell them a true story.

VIDEO RECOMMENDATIONS

We often hear people talk about how perfect nature is. How can all of this beauty around us be the result of evolution? Nature is so amazing that it can appear to be designed. Nature is awe-inspiring, I agree. I have hiked on all seven continents. From my campsite in the Australian Outback, I saw a dingo peering through the darkness. In the Amazon, I saw pink dolphins playing in the river. A wolf running through the woods in Yellowstone, a humpback swimming with her calf in Antarctica, a cheetah resting with her five cubs in a South African reserve. These moments have brought tears to my eyes. In perhaps my most powerful memory of nature, I witnessed a tiger stalking a deer in Ranthambore National Park in northern India. Believe me, I get it.

However, I am also very mindful of the many painful or strange examples we can find in the natural world. Parasites kill millions every day. Many people and animals are crippled by birth defects. Other examples of nature's imperfection are more odd than heartbreaking. My favorite evolution video illustrates that evolution does not make nature perfect; evolution makes nature just good enough, and it's full of mistakes. Evolution, as the narrator of the video says, has no foresight. The recurrent laryngeal nerve is a good example of this. This nerve runs from the brain to the vocal cords in all tetrapods. On one side of the head, it goes straight there, but on the other side, it goes down the neck, around a major artery of the heart, and back up again. It travels several feet out of its way. Why? The answer is in this video. It is a good way to show our students that evolution is not just all around them, all living things carry the history of evolution inside their own bodies as well.

"My dog Jinx got fleas. I purchased flea medication at my local pet store. Flea medicine is expensive! It cost me $78 for a three-month supply. Well, Jinx still had fleas days after I administered the medicine. I took her to the vet, and my vet casually said, 'Oh yeah, those meds do not work anymore.'"

I ask the students to tell me why these medications have been rendered ineffective over time. This short activity can be done orally or as a writing assignment, individually or in small groups.

It may be necessary to probe student understanding before the students reach the correct answer, especially if they have a limited understanding of genetics. Some questions a teacher can ask include:

1. Do you think every flea on my dog survived?
2. If not, why did some fleas die?
3. What did the fleas that survived have that the dead fleas did not have?
4. Did they pass this ability to survive, or resistance to the medication, on to their offspring?
5. What if they do not pass it on? Would it make a difference?
6. What happens to the flea population over time in terms of the ability to survive the medication? Why?

The answers I eventually receive sound something like this: "Some fleas were lucky enough to receive a random mutation which made them resistant to the medicine. Over time, the sensitive fleas died and the resistant fleas spread in the population." (One of the biggest misconceptions about evolution is that the entire process is random, just sheer luck. This leads people who do not understand evolution to bring up the fallacy, "Evolution is like a tornado ripping through a junkyard and creating a 747 airplane." I will discuss this more when I address misconceptions).

At a TIES workshop in Port St. Lucie, Florida, Vázquez demonstrates how to present a lesson in hominid skull comparison.

Once the students understand that some fleas got this lucky genetic mistake that helped them survive and passed it on to their offspring, I always say, "OK, we just covered evolution, let's move on to ecology!" After a few seconds of disbelief and/or laughter, I explain that natural selection, one of the primary mechanisms for evolution, happens when there is variation in a population (resistance vs. sensitivity to flea medication, for example). Random genetic mutations cause the variations. Sometimes, these random genetic mutations can be beneficial for the survival of certain individuals in the population. If this mutation is heritable (passed on to the offspring), the population can change over time as more offspring are born with the beneficial mutation.

Two excellent videos can follow this simple exercise. One is *The Evolution of Bacteria on a 'Mega-Plate' Petri Dish* and the other is *Excellent and Simple Explanation of Natural Selection,* both available on the TIES website.

The *Mega-Plate* video shows how a species of bacteria slowly move across a giant petri dish with increasingly stronger doses of antibiotic. It is the same concept as the flea medicine example. The *Simple Explanation* is a brilliant short video on an imaginary species which slowly changes over time to blend in with its environment. As the environment changes, other random mutations lead to more genetic changes in the species. Teachers can pause the video at specific points and discuss the important concepts using the following guided questions.

Time stamp 0:23

What similar traits do members of the species have? What different traits do members of the species have?

Time Stamp 0:46

Which members of the species are more vulnerable to the predator?

Which members of the species are more likely to survive?

Can you predict what the population will look like over time?

Time Stamp 0:52

Can you describe natural selection? Natural selection is the process that leads to the buildup of favorable characteristics in a species over time, because individuals with favorable traits survive to reproduce.

What is adaptation? The way a population changes over time to suit its environment. *Remember:* Individuals do not change over time, the population does.

Time Stamp 1:33

What will happen to the green individuals in this sandy environment? What if the environment changes and gets greener?

Time stamp 1:49

How does variation play a role in the survival of the species? (It gives the population a chance to survive when there are changes in the environment).

The bell-ringers and videos I have just mentioned serve more than one purpose. As mentioned earlier, they show students that the basic idea behind evolution is not difficult to understand, it just makes sense. Second, if their parents have reservations about evolution being taught in their children's science classes, they will certainly ask their children what they learned in class on the first day of the evolution unit. Who can disagree with flea medicine losing

its effectiveness, or the well-known concern that many antibiotics no longer protect us from once treatable bacterial infections? By introducing evolution with the issue of flea resistance, a teacher can eliminate (or at least attenuate) the reprisal of some parents.

Which leads me to one last thought about introducing your evolution unit in this manner. Evolutionary biologists do not distinguish between microevolution and macroevolution. I learned those terms when I started researching creationist arguments against evolution. Microevolution refers to changes *within* a species. Macroevolution refers to how a species can become an entirely separate, distinct species. These activities both fall under what creationists would call microevolution. They may argue that while so-called microevolution is not controversial (again, how in the world can anyone deny flea or bacterial resistance over time), macroevolution has never been seen. This is patently false. Speciation, as biologists call it, is the phenomena of one species changing over time into a separate species. Speciation has been documented all over the world. A simple Google search will lead to hundreds of examples.

Scientific American covers the topic well in "Evolution: Watching Speciation Occur."

Favorite Investigative Unit or Activity

"This Lab is for the Birds" is an active learning lesson I created based on the work of Peter and Rosemary Grant, two biologists who have spent decades studying the medium ground finches of Daphne Major in the Galápagos Islands. The Grants have meticulously measured the beak depth of this species and how it has changed over time depending on food availability.

"This Lab is for the Birds" places students into groups of four. Each student in the group has a different beak phenotype: fork, spoon, knife, and toothpick. In front of each group is a plate with raisins. A timer tells the students when to start collecting raisins one by one using their utensil. When the timer stops, students count the raisins they have each collected. After two more trials, the raisins are replaced with uncooked popcorn kernels. This change represents the change that took place on

Daphne Major during several years of severe drought. As one would expect, the toothpick and knife phenotypes do not do very well collecting popcorn kernels.

The Grants found that the medium ground finches with narrower beaks did not survive during the years of severe drought, and the entire population moved toward thicker beaks. This activity requires that the students collect and analyze data, drawing conclusions from what they observe. I created a video of my own students trying out this investigation to help teachers who have never tried it before. It is available for viewing along with the lesson itself.

HHMI's BioInteractive website has an excellent short film featuring the Grants that teachers can use to supplement the activity. Please note: many textbooks and curriculum guides contain a similar activity about adaptation. Students use similar utensils and capture food much like this activity. There is one main difference. In the activity on adaptation, the different utensils represent *different* species. In this activity, the different utensils represent different phenotypes *within* a species. Natural selection occurs within a species.

If you are really short on time, The TIES Time Machine is an engaging online game in which students choose a "starter" population of imaginary animals. This population then goes through a series of environmental changes over time. The goal is to succeed in getting your population to survive a million years. Students quickly learn that making sure that their starter population has a variety of traits is helpful. They also learn that some environmental changes, such as a volcanic explosion, will doom their species no matter what.

Dealing with Conflicts

When I described my favorite bell-ringer, I mentioned that some students think evolution is only about random mutations. How can a bunch of random mutations produce the vast array of life we see around us today?

Scientists and authors have repeatedly explained the reason that this does not accurately describe evolution. The explanation I use to help students understand this came from a *New Scientist* article in 2008, "Evolution Myths: Evolution is just so Unlikely."

In a recent TV special shown in the UK, called *The System,* a mother with big debts was persuaded to borrow even more money to bet on a horse race. Having been sent correct predictions of six previous races, she believed illusionist Derren Brown really had come up with a foolproof system for predicting the outcome of races.

In fact, the producers of the show started by sending different predictions to nearly 8,000 people. After each race, those sent predictions that turned out to be wrong were eliminated and another set of varying predictions sent to the remaining participants. What appears utterly extraordinary at first—sending someone correct predictions of the winners of six races—seems very ordinary as soon as you understand that thousands of people got wrong predictions.

Confronted by the marvels of the living world, many people jump to the same conclusion to the woman in the program: they cannot be the result of chance alone. But what we don't see are all the failures: the countless numbers of creatures that died in the egg or in the womb, or hatched or were born with terrible defects, or fell victim to predators or disease because of some weakness.

In the wild, most individuals die long before they get a chance to reproduce. The living organisms on Earth are the result not just of six rounds of selection, as in the TV program, but of trillions. This, not chance, is the crucial factor in evolution.

2

The Accidental Evolution Teacher

Nikki Chambers, MS in Biology
Southern California

I was going to be a journalist. Fifteen-year-old me was quite convinced of this. Science was for nerds, and I was having enough trouble fitting in socially in high school without adding to the problem. Science was also for *boys* (or at the very least for girls whom boys would *never ever look at*)—or at least, that was the undercurrent message I got from my formidable, very "ladylike" mother in the mid-1970s. In hindsight, the How and Why Wonder Books so enticingly displayed at the checkout counter of the local grocery store, which I coveted and which my mother would occasionally buy me, might have been a hint of the inevitability of things to come. More than forty years later, the two titles that I remember best—indeed, the issues I memorized forward and backward—were *Planets and Interplanetary Travel* and *Oceanography*. Throughout my childhood, my voracious consumption of books included science fiction titles such as *A Wrinkle in Time* and *The Wonderful Flight to the Mushroom Planet,* but that might have also been shaped by my personal experience as the daughter of my Apollo scientist father. One of my earliest memories with Daddy was of him taking me to the spacecraft assembly facility in Long Beach, California, to see a shiny, beautiful cone-shaped space capsule in which three people—people!—were going to travel *to the moon.* When Neil Armstrong took that "one small step for mankind" on July 20, 1969, it was the day after my ninth birthday, so surely it was my beloved father's birthday gift to his starstruck daughter, who could tell you by heart how many moons every planet in our solar system had (thanks, of course, to the relevant How and Why Wonder Book).

At fifteen, though, the stars were obscured by more pressing worries, like how to make friends at a huge new high school, and why people laughed at the way I dressed, and how to get people to be interested in me for more than just copying my homework. I still loved books and words; English class was a joy, as were my foreign language classes. History was all about old, dead white guys and battles and dates—yuck. Geometry was cool, but trigonometry whizzed right past me as I spent most of the class passing notes about much more pressing issues to my best friend across the room. And science was a necessary evil, a two-year graduation requirement that got put off until my junior year. The cool, popular, fashionable girl that I yearned to be hated science . . . right?

So, uncool, unpopular (except for homework copying purposes), unfashionable, fifteen-year-old me plopped down in Mr. Calabrese's Introductory Biology class at the start of her junior year. Suddenly, in the words of Madeleine L'Engle, "A straight line is not the shortest distance between two points." Memorably and truly life-changingly, Mr. Calabrese opened my eyes to the endless fascination of life itself. He recommended an optional study resource, whose title I don't remember, but it was an absurdly fat paperback that I bought at ASB while other people were buying tickets to the homecoming dance. I devoured it while sitting on the curb by myself while those other people laughed together at the lunch tables and copied each other's homework. I immersed myself in kingdom-phylum-class-order-family-genus-species and cell parts and *all* the steps of the Krebs Cycle. And, best of all, I sought to understand how one cell becomes two; and two become four; and four, in days/weeks/months, become something wonderful, something amazing, something almost miraculous: a *new being*, something that had never before been seen in exactly this particular incarnation, something that grows, and changes, and is shaped by both the message within and the world it inhabits. That *being*, whether it is a bird or a fish or a slime mold or a human, will—if it is lucky—pass on the essence of its self to yet another being, that will be the same but different. We, in the late 1970s, were only starting to understand those details, to marvel at life's grand choreography.

Then, of course, came the panic. The journalistic goal crumbled to dust. I needed to learn all the intricacies of life, particularly of that marvelous biological machine, the human body, and here I was, with only one more year in high school to play science catch-up! There was no alternative: I took chemistry with the happily explosion-inclined Mr. Ryan over the summer, then took both physics and AP biology during my senior year. Extra math courses were crammed in as well, and before I knew it, I was just twenty-one years old, with both bachelor's and master's degrees in biology from UCLA, walking down the steps of Young Hall after my last exam with tears streaming down my cheeks. I had adored every science class I took. I soaked in the intricacies of chemistry as taught by the masterful Paul Boyer, who just a few years later would receive the Nobel nod for teasing out the secrets of mitochondrial function taught a biochemistry class. An embryology class was led by a professor whose English was so bad that I'm still not sure how she communicated the amazing phenomenon of gastrulation to me. A field ecology class remains memorable for being punctuated by the terrified screams of a fellow student who woke up during one of our overnight camping trips to find a tarantula parked serenely on his chest. I had loved it all, I had finished at such a terribly young age, and all I could think, as I traipsed dejectedly to the parking structure

to make my last trip home, was . . . *now what?* What do I do next, now that the learning is done?

Not unlike the girl who was *sure* she would never be a scientist, if you had asked me that day (or, indeed, on any of the days for almost the next twenty years) whether I might consider teaching, I would have laughed you out of the room. Teenagers were *awful*. They were scary, they teased you, they did dumb things that weren't really very funny, and they cheated and copied your homework. Why on earth would I want to purposefully spend my days with such a horrible segment of humanity?

I spent most of the next two decades doing, well, other things. Among them were two nine-month stretches in which my body became a vessel for one cell that became two, and two became four, and days/weeks/months later, the miracle of life burst forth from *me*. As my children grew, I read them books about everything from planets to oceans, from stars to little dogs to giving trees to hobbits. I told stories to those two remarkable little beings who were shaped by the billion-year-old chemical messages that were gifted to them from their father and me, two halves made into wholes, unique in all the history of the universe.

Life, it has been said, is what happens while you are busy making other plans. Life also has a sneaky way of weaving together threads that you don't think will ever connect. Through a series of events that happened over the course of those two decades, the child who loved oceans and planets and stories, the girl who was sure Daddy had given her the moon landing, the awkward teen who was more at home in a eukaryotic cell than at a pep rally, the young woman who laughed at the suggestion that she become a teacher, and the mother who told tales about dragons and dinosaurs once again became a student, enrolling in Cal State Long Beach's science education program. Sharing the house with adolescents who weren't really all *that* bad, I thought the idea of passing on the love of the universe and all it contains developed an irresistible appeal. In a flash, it seemed, I was once again in cap and gown, and then fortuitously hired at my children's high school. (My son, who was about to start his freshman year, was, of course, appalled that his mother should thus ruin his life. We struck a deal: I wouldn't embarrass him if he wouldn't embarrass me. It worked. Mostly. He was

VIDEO RECOMMENDATIONS

It's hard to go wrong with *Making of the Fittest: Natural Selection and Adaptation.* The explanation of evolution from the ecosystem down to the molecular level in eleven minutes, complete with cute mice and scary predators, never fails to capture kids' attention and make them go, "OK, I get it!"

multitalented, handsome, and popular; I set things on fire and threw things off the balcony. It was a pretty good deal for both of us, in the end.)

I taught our school district's version of integrated science during those early years. I covered a little bit of everything, from astronomy to zoology. None of my students ever knew (I don't think) that I madly crammed our units on seismology and meteorology the night before since I'd never taken any classes on these topics. But in those lessons on things I had long known and loved—including the ocean deep, or the moons of Saturn (which, by 2003, numbered considerably more than the nine I had memorized from my How and Why Wonder Book)—I fell in love all over again.

I wasn't the only one. Two topics consistently captured my students' imagination: astronomy and human paleontology. Astronomy received a scant three weeks of time in the curriculum; human origins, none at all. Evolution, my colleagues suggested, was best taught after "The Test" (a.k.a. the old California STAR test), since there weren't very many relevant questions on it anyway and we had to cover the checklist of "students should know" factoids called out so systematically in the state standards. My students, however, kept asking questions. Where did we come from? How did it all get to be the way it is? Are we alone? Are we unique? Are we connected? Somehow what we were teaching didn't align all that well with what they wanted to know—or, honestly, what I really wanted to teach. I wanted to answer their questions, those deep, ancient questions about our place in time and space, things that are at the core of what it means to be human.

In October 2005, I attended a workshop taught by my inspiring colleague and friend from the SETI

Institute, Pamela Harman, at the California Science Conference. The workshop showcased a Human Evolution unit, one of six units from the SETI Institute's "Voyages Through Time" high school integrated science curriculum. Four of the other five units had the word "evolution" in them: "cosmic evolution," "planetary evolution," "evolution of life," "evolution of technology." The sixth unit (albeit third in the instructional sequence) was equally tantalizing: "origin of life." To say I was hooked is putting it mildly. I remember coming back to my vice principal, Pam Metz, tripping all over myself with excitement, telling her that somehow—in spite of the standards!—I needed to teach a class like this. I needed to tell students these stories as I took them on the ultimate journey through time. Bless Pam for not even blinking twice before encouraging me to make it happen. And it was thus that I became, unintentionally, somewhat accidentally, but absolutely unequivocally, an evolution teacher.

I spent the next two years planning, writing, revising, submitting for approval, even doing battle with a district administrator who sniffed that this was a history class, not a science class. In 2006, I attended a three-day conference in Venice, Italy, simply entitled Evolution, with one day each spent on the evolution of the cosmos, the evolution of life, and the evolution of the human mind—sponsored by, among others, the Catholic Church. Finally, in September 2007, I welcomed my first twenty-five high school seniors to a course simply entitled Astrobiology, which I hoped would be a survey course about the evolution of, well, *everything!* You cannot truly understand the nature of an individual, an ecosystem, a planet, a solar system, a galaxy, or a universe without understanding how it got to be that way. As Carl Sagan

BOOK RECOMMENDATIONS

Zoobiquity: The Astonishing Connection Between Human and Animal Health (2013), by Barbara Natterson-Horowitz is a fascinating and absolutely delightful read about the commonality between humans and all other life forms. Natterson-Horowitz, a brilliant cardiologist, is a cardiology professor at the David Geffen School of Medicine at UCLA and serves on the medical advisory board of the Los Angeles Zoo as a cardiovascular consultant.

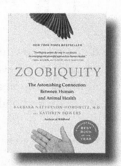

said, "If you wish to make an apple pie from scratch, you must first invent the universe."

In hindsight, the hubris of such an undertaking surely reflects some combination of unbridled passion, stubborn persistence, naivete, foolishness, and a star-struck sense of wonder. Twelve years in, all of the above still apply, with the "sense of wonder" consistently winning out. The more you understand how things became the way they are, the more you hunger for what you don't yet know, and the more all this thrilling knowledge simply has to be passed on to others.

In the early years of the course, I struggled most with the same topics that I had been least prepared to teach as a new teacher: planetary geology, ocean-ography, and meteorology. With every new school year/month/unit/lesson, I am challenged to share cutting-edge knowledge in ever-developing fields with my students, in ways that they can understand and embrace. I could not possibly have been successful without an army of amazing and supportive colleagues at world-class institutions like the SETI Institute, the Jet Propulsion Laboratory, the NASA Astrobiology Institute, the Howard Hughes Medical Institute, the astrobiology departments/centers at the University of Edinburgh and the University of Hawaii, and both the Department of Integrative Biology and the Museum of Paleontology at UC Berkeley. The latter conducts a five-day Think Evolution workshop for teachers every summer, at which I had the great good fortune to meet Bertha Vázquez, whose passion for outreach complements my own. Bertha introduced me to her brainchild, the Teacher Institute for Evolutionary Science (TIES). This insightfully developed hybrid of virtual and actual professional development opportunities and teaching/learning resources has allowed me to extend my passion for teaching evolution to other teachers as well. Teaching evolution, as all of us who do it are aware, may be rich and fascinating, but also has layers of nuance and complexity that require help if they're to be taught correctly. And of course, teaching evolution faces unique challenges in today's society—challenges that are much easier to overcome when supported by a robust platform of resources.

After trying out the TIES curriculum in my own classroom, I quickly realized that these resources would be a welcome gift to my district's middle school teachers, who were in the midst of the "now what?" phase that accompanied their shift to the Middle School Integrated Model for implementation of the Next Generation Science Standards. Suddenly we had both teachers who had never taught life science before, and teachers who had minimal preparation for teaching evolution in the richer, more meaningful way expected by the NGSS. Presenting a TIES professional development workshop to all of our eighth-grade science teachers was a perfect solution to the problem (and, frankly, almost too easy, given the well-structured nature of the presentation resources themselves). The two ultimate questions, though, would remain unanswered until the current school year: would the middle school teachers actually use the resources, and would students demonstrate long-lasting understanding?

The answer to both questions, to my delight, has been "yes." My current group of high school introductory biology students has clearly experienced the TIES lessons in middle school, and I love knowing that I can build upon that learning to offer them a deeper dive into biological evolution than I have ever been able to do in high school before. Two years from now, when some of those same students enroll in my Astrobiology class as high school seniors, we will be able to explore those quintessential human questions—"Where did I come from? How did I get here? Are we alone?"—more robustly than ever.

I tell all my students, both my freshmen and my seniors, that years from now, I want them to visit me and tell me that, no matter where their paths have taken them, they have never forgotten two key phrases for all evolutionary thinkers and learners: "I wonder . . . ?", and "What's the evidence . . . ?" Richard Dawkins has written eloquently of the *Magic of Reality*. Understanding the ancient, grand evolutionary processes that have made us part of that reality is truly one of the most marvelous things there is. It has been the happiest of accidents, and it continues to be a privilege and an ongoing joy to impart that wonder to those who will listen.

Introducing Evolution

My high school biology class is based on anchor phenomena and the stories that connect them. As a result, I don't have a single defining "bell-ringer," nor do I have a "first lesson" (or a "last lesson") on evolution, or indeed a defined "evolution

Nikki Chambers with two young evolutionary biologists discussing the evidence for our own evolution by analyzing and evaluating cast models of hominid and non-hominid primate skulls.

unit." Our studies of evolution emerge naturally from the ecological investigation with which we open our school year.

 I teach in a school on the Southern California coast; all of my students spend their summers at the beach, with kelp between their toes. Little do they know, at first, what an amazing hundred-year story of diversity, destruction, and rebirth has happened in the waters twenty minutes from our classroom. After building their understanding of a rich and complex ecosystem, I ask a deceptively simple question: if this has been the story of the last hundred years of the kelp forest, how much deeper does the story go? Just how long has the kelp forest been there? What was there *before* there was a kelp forest? How far back does *that* story go? Was there always a kelp forest, all the way back at Earth's beginning? From this, opportunities to tell the story of the coevolution of Earth and life naturally emerge.

If I had to identify an activity that focuses my students' sense of wonder about the processes of evolution early on, it would be an activity I developed, based on the Howard Hughes Medical Institute's excellent poster, "Human Evolution Within the Tree of Life." The activity, which takes about thirty minutes, includes both collaborative discussion among student groups, as well as individual written responses to focus questions.

Materials: HHMI "Bones, Stones, and Genes" poster (one per table) and poster questions (one per student)

Explore: HHMI poster exploration

- Provide each table with one poster.
- Distribute question handouts, one per student.
- Students work as a table group to explore and discuss poster.
- Each student responds individually in writing to the poster questions.
- Students turn in their answers when they finish.

Favorite Investigative Unit or Activity

I would do the "Biogeography of the Malay Archipelago" lesson that I based on several HHMI resources. HHMI's page of resources is linked here, and the lesson follows.

Dealing with Conflicts

I frequently have to deal with students and families who struggle with a perceived conflict between faith and science as mutually exclusive ways of knowing. I purposefully engage my students in conversations about the importance of different ways of knowing. I find the Smithsonian Museum of Natural History's "Cultural and Religious Sensitivity Teaching Strategies Resource" to be extremely helpful, as well as the teaching resources from the National Center for Science Education.

3 The Cherokee Creation Myth

Amanda Clapp, MEd
North Carolina

The classroom is silent. It's dark. Behind us, the lists we have generated on the interactive white board lend an eerie glow. The hush of the crowd at the beginning of a story is a palpable thing. There's a tense pause, then we dive together into a story of creation.

The earth is a great island floating in a sea of water, and suspended at each of the cardinal points by a cord hanging down from the sky vault. When the world grows old and worn out, the people will die and the cords will break and let the earth sink down into the ocean, and all will be water again. We are afraid of this. When all was water, the animals were above in Galunlati, the sky arch, but it was very crowded, and they were wanting more room. They wondered what was beyond the water, and at last Dayunishi, "Beaver's Grandchild" the little water beetle, offered to go and see if it could learn. It darted about in every direction over the surface of the water, but could find no place to rest. Then it dived to the bottom and came up with some soft mud, which began to grow and spread on every side until it became the island which we call the earth. It was afterwards fastened to the sky with four cords, but no one remembers who did this. At first the earth was flat and very soft and wet. The animals were anxious to get down. At last they sent out the buzzard and told him to go and make ready for them. This was the Great Buzzard, the father of the buzzards we see now. He flew all over the earth, low down near the ground, and it was still soft. When he reached the Cherokee country, he was very tired, and his wings began to flap and strike the ground, and whenever they struck the earth there was a valley, and where they turned up again was a mountain. When the animals above saw this, they were afraid that the whole world would be mountains, so they called him back. But the Cherokee country remains full of mountains to this day.

Swimmer, Ayuini (Duncan and Riggs 2003).

A collective sigh goes up. We flip the lights back on. "So," I ask. "Which set of explanations does our story support? Turn and talk." The students are buzzing, laughing, and talking about their lists and the story. The lists are explanations of the diversity we discovered in the outdoor classroom, and the students collect the explanations into those that can be supported by the Cherokee creation story. Our school is in the Eastern Band of Cherokee Indian homeland, and many of the students have ties to the culture and traditions of the tribe. After five minutes, the students share their analysis: some explanations that can be supported through traditional stories are: "Diversity just exists. It always has." And "God." And "Aliens." And "I don't know." The explanations for diversity that the students did not include were "evolution," "change over time," and "natural selection." This is going to be fun.

A Journey Toward Education

Growing up on a wildlife sanctuary in Massachusetts, I loved to learn. My childhood was filled with nature; I loved school and followed all the rules in my suburban system. After school, and in the summer, I was in the woods, climbing trees, attending nature programs at the sanctuary, and learning about nature the way young children learn a second language. Nature was my window to human diversity as well; my family traveled to see wildlife, and I could meet people from everywhere in the world, communicating despite language differences, enjoying our similarities. Based on my interests, of course I would continue my education to become a scientist. As an undergraduate, I discovered wildlife biology and physical anthropology, and I was able to pursue an interest in nonhuman primate ecology and conservation. Pretty much, lemurs were awesome, and Madagascar needed help with conservation.

My travels after college, working as a field assistant in Madagascar and as a naturalist in a number of places, confirmed my plan; I went to graduate school to pursue evolution and ecology. This is how I ended up in front of an undergraduate class teaching a physical anthropology lab. It was the class that undergraduates could take to get that pesky science requirement out of the way. Many students told me they just weren't good at science, or they just didn't like science. I discovered that these college students had missed huge chunks of basic biology; I was teaching them about cells, about genetics, about adaptations. They had somehow missed these foundational concepts in their K-12 education. I realized I was needed elsewhere: I went to teach middle school.

Since life is complex and wonderful, I ended up in Southern Appalachia. Western North Carolina is a tourist destination in the summer—the rivers are filled with kayaks and fly fishermen; the roads and trails of the Great Smoky Mountains National Park are packed with hikers and photographers. Giant campers lumber up and down the steep roads. While it is a vacationer's paradise, this area remains socioeconomically depressed, and many families live month to month in traditional mountain homesteads. The students I have been teaching for fifteen years are diverse; adventurers' children, preppers' children, professors' children. While some students miss school in the fall to cut Christmas trees on the family farm, other students miss school for a cruise or a ski trip. One of the significant things that ties them together is their connection to the land. This connection is strong. More than half the students I teach go home to the woods and the garden after school, to run with their friends and dogs or to hike and fish. Of course, this means we learn outside. We use the outdoor classroom to do citizen science, setting up our own methods for testing hypotheses about growing seeds and the birds to expect at our feeders. In many ways, I have the perfect situation for teaching evolution.

How Do We Learn About Evolution?

 My students get excited about iNaturalist. It's an app that allows us to catalog every living thing we see on our campus and in the outdoor classroom. The eighth graders went out to document as many things as they could, and they came up with one hundred living things in ten square meters. When we subtracted selfies and too-close friend pictures, we were at eighty-nine. At this point, we could start the conversation; why is there more than one kind of grass? Tree? Insect? My students were perplexed. They couldn't explain why there was such a diversity of things that didn't seem to matter. "But we don't need all the little bugs—they just go in our eyes!" They argued. Of course they're an inconvenience, but in that same spot, we saw Eastern Phoebes, flycatchers that exclusively eat those little bugs. We also saw Tree Swallows—same thing! And we saw toad and frog tadpoles, who survive on midge larvae, in the tiny pond. Without those annoying insects, what would happen to this system? We could look at these bugs that get in our eyes more closely, and we can find them as integral parts of both aquatic and terrestrial ecosystems. Thinking about the diversity of animals at different trophic levels was also a good way to talk about the fact that individual species have different adaptations for the same thing; we watched the Phoebe

VIDEO RECOMMENDATIONS

A video clip about Alfred Russel Wallace from *The New York Times* fills in some historical context about defining evolution in the 1800s. The animation is a brief biography of how Wallace came to the same conclusions as Darwin about natural selection at about the same time in history. His was more of an underdog experience, and the students can talk about the scientific and social changes that made travel, learning, and communication possible.

perch, bobbing its tail, then fly out for a snack and return to its perch. On the other hand, we never saw the Tree Swallows land. Even though they have similar food, they have a different method of eating.

We made lists in the outdoor classroom. These lists included ideas about how all those different organisms ended up in our campus, how they ended up depending on the other organisms there, and how they have been so successful. The lists were also full of the ideas that explain why the diversity exists. When we go back to the classroom, there is so much to talk about. We pause, listen to the Cherokee story, and then break out our lists. Our conversation is brief; we talk about ourselves. People love to explain what we see; the explanations that are supported with evidence are the ones that explain the diversity of life and how organisms change over time. Other stories have cultural significance, but they have no proof.

Soon, students are researching. They choose a domesticated organism and try to trace it back to its pre-human form. In two days, the students have constructed enough knowledge that we can apply some vocabulary and move on. At this point, I like to show the Howard Hughes Medical Institute's BioInteractive video on teosinte, the grass that early Native Americans domesticated to become corn (HHMI n.d.). There's more than one documentary on dogs, but half the class usually chooses dogs, so I don't worry about showing them. The product the students make is a simple flow chart, with drawings or pictures of the different variations in their species. We can choose any of these flow charts to teach how

humans have selected for certain traits. This is an easy concept for my students, because in our area, hunting dogs are actively bred, and there is enough agriculture that students have seen different breeds of cows, goats, and chickens.

The next step is for us to investigate natural selection, and we have a wonderful time. We use the TIES peppered moth simulation, which is easy and fun, to introduce the idea of predation as a selection pressure. We play Project WILD's Oh Deer! game, adding and taking away traits depending on limiting factors. We also spend a day with young Darwin on the *HMS Beagle,* discovering fossils in Argentina and all the island biogeography of the Galápagos. We draw in the examples from other topics we've studied, and the students' list of evidence for evolution grows. By the end of our unit, the students have acted like scientists by making observations, hypothesizing, doing research, and connecting their research with a scientific theory: evolution.

Why is Evolution Important in Middle School?
Young adolescents are on the cusp of so many things. They want to be adults and children at the same time; they want to make difficult choices, but they need the strategies to navigate their choices. Middle school provides the opportunity for individual thought, a space and a learning community that lets them begin to evaluate both the facts they packed from their childhood and the

Amanda Clapp works with students to identify terrestrial macroinvertebrates in the outdoor classroom. Students use shaker boxes to separate the invertebrates from the leaf litter, then post their discoveries on iNaturalist (https://www.inaturalist.org/).

facts that are presented as they enter adulthood. What studying evolution has done for my students is provide them a way to establish their own understandings about the world. In the rural South, the church is a foundation of community, a way for neighbors to keep in touch, and a way to lend a hand. Here in the mountains, communities have been isolated from each other, and churches have been a way of forming and maintaining connections. In many churches in my region, evolution is misrepresented and sometimes proselytized against. As a result, many students who arrive in my classroom have only heard the word from the pulpit, and in a negative light. The opportunity for students to develop ideas separate from their church is a powerful one. I provide time for the students to do their research, collect evidence, and work through the ideas that as a scientist I took for granted. While I provide the Cherokee creation story as a way for students to organize their thoughts, I only provide opportunities for evidence the rest of our learning time. Early in the process, we do the "What is a Theory?" probe (Keeley et.al. 2008) which helps the students frame their understanding. An important thing for us to remember is that

BOOK RECOMMENDATIONS

For teachers who are interested in Darwin's writing, part of what makes *On the Origin of Species* exciting is its historical context. In the mid-to-late 1800s, an amazing time for thinkers, the microscope and other technology were becoming better and more accessible. People were becoming more able to travel the world and see how nature—and people—that are thousands of miles away can be similar to what they see at home. Both books take the original text Darwin wrote and add modern examples and support. The first, *On the Origin of Species: The Illustrated Edition* (2011) is illustrated by David Quammen, who's a great writer in his own right. It's an easier read for a teacher, and a few pages could be used as an example with a class. The second book recommendation is *The Annotated Origin* (2011), in which Jim Costa gives lots of examples from modern biology to support Darwin's observations and hypotheses.

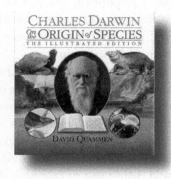

the scientific theory is—as yet—not disproven. We spend some time thinking about science as a self-correcting process; the more we learn, the more we add to the body of knowledge that supports theories.

As students leave elementary schools, they want to make their own choices. It's extremely important to teach them how to vet sources and consume media in a way that allows them to analyze the evidence and make their own choices. Providing a format for the students has been helpful for me. I use the claim-evidence-reasoning approach with students. You can make a claim, but it has to be backed up with a lot of evidence for the rest of us to consider it. Evolution helps us put this process into practice; if you make a claim and it is supported by data, it can be considered. The amount of evidence the students collect in our research allows them to make a claim and discuss its validity. I hope this practice then transfers with them to the adult world, where they consume news media, social media, and they are still intentional, productive members of our global society.

Recommendations for Teaching Evolution in Middle School

1. Embed it. Evolution is the basic paradigm of life sciences, so don't wait until the dreaded "Evolution Unit." I currently teach sixth, seventh, and eighth grade at a university lab school, and we have talked about evolution in all three grades by the fourth week of school. When I teach natural selection directly, the students will have examples from the rest of their year to support it. North Carolina has not yet adopted the Next Generation Science Standards (NGSS), but they are pretty similar, and using the *Frameworks for Science Education* is helpful (NRC 2012). This gives me an opportunity to teach general science and embed evolution everywhere. In sixth grade, we talk about adaptations of flowering plants. Just the fact that flowering plants evolved toward the end of the dinosaurs' reign gets everyone's attention. The alignment of structure and function in photosynthesis and reproduction has us shouting about coevolution like banshees. The seventh grade covers human body systems. We spend a lot of time talking about why we can dissect the organs of other vertebrates to infer the structure and function of our own systems—we all use the same plan, just modify it as needed. We even read an excerpt from *Your Inner Fish* (Shubin 2008). In eighth grade, we study viruses and bacteria, so students have multiple lines of evidence supporting evolution within populations of humans

and populations of microbes. I had parents call and ask if I told their kids to get flu shots. Of course not, I responded. We just talked about how vaccinations work, and we calculated the number of viral generations in one year. The students made their own connections.

2. Make it relevant. In middle school, students are constructing identity. We can use our discussion of evolution to support human rights, and to embrace the similarities between ourselves and our friends. I use a TED talk by Nina Jablonski (2009) or a TED Ed lesson by Angela Flynn (2016) to compare the cultural definitions of race with the scientific fact that race simply doesn't exist. We talk about evolutionary fitness: I challenge the toughest, biggest student to an arm wrestle, and I lose. But, I tell them, I am more fit! I have two children! Boom! Evolution is one of the best ways to talk about why polar bears can't survive climate change, or why the next outbreak of Ebola or another disease could be right around the corner.

3. Find your people. Many middle school educators are islands. We are often surrounded by a team of great educators, yet we are the only science-oriented folks in the hall. To develop pedagogy and to deepen our content, we need to find each other and develop our own networks of science educators. It is within these communities that we can grow. On the other side of the coin, include your middle school team in an evolution-oriented unit. An easy entry into this collaboration is this: my team and I planned an interdisciplinary unit that incorporates microbes and epidemics research in science and math, a novel study on *Fever 1793* (Anderson 2000) in ELA, and the introduction of smallpox to North America in social studies. Your people are also the scientists who are informed by evolution every day. Reach out to your nearest university, college, or community college. I once celebrated Darwin Day by developing a program for my students at our local university. I asked the biology professors if they could give fifteen-minute presentations about an evolutionary aspect of their research. They all agreed, and the evening was amazing. My students learned about why chickens squawk at (almost) every falling leaf, how freshwater mussels mimic breeding fish to reproduce, about the coevolution of humans and our gut flora, and about spider predatory adaptations. The community connections and content enrichment are a great combination.

I love being a teacher. Every day is an opportunity for learning, and my students teach me about mindset, empathy, and grit every day. I have come to see my job in a different light from when I entered the profession years ago; my job is to provide the students with the time and the tools to develop their skills. One of the most important skills they can develop in a world overrun with misleading marketing, unkind tweets, and a severe social media presence is critical thinking. Evolution, with its many layers of evidence but its slightly controversial regional connotation, is one of the best ideas I can use to prepare my students for the lives ahead of them. The social and scientific history of evolution is fascinating, and investigating the sheer volume of evidence that supports a claim gives my students the chance to practice being discriminating readers, careful thinkers, and citizens who respond to the world with curiosity and constructive work. Evolution is not just a tenet of science, it's a way to view the world with appreciation, and to see the connections with all things. In middle school, connecting students with this way of thinking prepares them for the rest of their lives.

Introducing Evolution

Since evolution is the paradigm that organizes most of life science, I don't introduce it. I teach in North Carolina, and the state curriculum mirrors the Next Generation Science Standards, so when I teach sixth grade, we talk about plant adaptations for reproduction and growth. In seventh grade, we talk about protist and human adaptations for life. In eighth grade, the standards include biological evolution, but my students are not surprised by the concepts, even as they tell me they don't believe in the vocabulary.

Middle school students are the smartest humans around; they navigate complex social systems while working to keep their families and their teachers happy. They are discovering new ways to interpret the world that are based on their families and culture, but also based on

Students use sweep nets to find insects in the outdoor classroom. They collect ants, wasps, bees, and grasshoppers, take pictures, and post their discoveries on iNaturalist. This backyard bioblitz, and inventory of all the living things in the schoolyard, is the introductory activity for a middle school unit on biodiversity and evolution.

their access to media and learning separate from their childhood influences. They may seem distracted to an adult in any one situation, but balancing all these tasks can lead to distraction.

Many of us have misconceptions about science based on popular media, on our personal experiences, on our lives. In class, as my students embark on their inquiries of any topic, I try to assess their misconceptions, and correct them as part of my responsive classroom. One misconception many people, including my students, have is

that "evolution is just a theory." To tease that meaning apart, I use a formative assessment probe developed by Page Keeley et. al. (2008) in the *Uncovering Student Ideas in Science* series. Students are asked to identify the attributes of a scientific theory, and after they have chosen from the list of ten characteristics, they must organize their evidence and present it.

I have used this activity in two ways in the classroom: First, I have used the claims-evidence-reasoning format and asked students to define what a theory is, then support it using a formal claims-evidence-reasoning writing piece. Second, to look at classroom trends, I ask students to participate in a sticky-note histogram on the wall. I put the ten assertions along the bottom, and students add a sticky note above each assertion with which they agree, forming a bar graph with more sticky notes above the characteristics of a theory with which most students agree. Assessing students' misconceptions this way allows me to develop targeted instruction to correct different misconceptions.

If I need to correct misconceptions quickly, or if the sticky note bars haven't changed for a day, I show this brief *Facts vs. Theory* clip from the TIES website.

Favorite Investigative Unit or Activity

One of my favorite activities to use is the peppered moth simulation. It is easy to set up and quick to clean. In an inquiry- oriented classroom, students can use it to collect data and then to ask different questions about the relationship between the environment and the population. The simplicity of the scenario makes it a fantastic performance task after other inquiry or a great introduction to natural selection. As my only activity, it would give students a taste of biological response in a population, but its simplicity also paints a too-neat picture of how predator- driven natural selection occurs. If I were to use this example for my students, I may introduce other factors in later rounds, including pesticides and color-blind predators.

Dealing with Conflicts

I live and teach in a very Christian community; the most common denomination is Southern Baptist, but my students belong to a number of conservative communities. When students tell me they don't believe in evolution, I simply agree. Belief is based on their faith, for which they need no evidence. It's not necessary. Evolution is a theory that is defined exclusively by evidence. Their faith and the evidence for evolution simply do not overlap. As we learn together, some students remain couched in their dogma, while others begin to understand the space between science and faith. At the end of class a few years ago, a girl remained seated. I asked her what was wrong, and she looked up at me with tears in her eyes. "This just makes sense," she said, "but it's different from what I was taught before." That's the great thing about middle school; students can collect information and define their beliefs and understandings based on more than one perspective.

4 Showing Evidence of Evolution

Kenny Coogan, MA
Florida

I completed middle school and high school with very little cognitive strain. My freshmen year in college, I was set to be a biology major. Receiving a D in an Intro to Evolutionary Science class my first semester was a shock to my ego. I was so distraught and confused the following semester I changed my major to animal behavior. When I graduated three years later, I started working for zoos and aquariums in their education departments. I learned what I needed from my peers on the job.

Ten years after working for zoological organizations, I became a middle school science teacher. After all those years of focusing on zoo-themed science, I was assigned to teach topics ranging from electromagnetic spectrum to geography to my adversary of the past: evolution.

My first-year teaching was successful because of my district-assigned mentor. She once suggested that she was able to help collate, cut, copy, and preserve my lessons—and I took her up on that every week. While other teachers were spending time photocopying or other mundane tasks, I was researching lessons and grant opportunities. At the end of the year, I applied for the Beginning Science Teacher of the Year for the state of Florida.

I received the honor at the Florida Association of Science Teachers (FAST) conference in Tallahassee. Before my award at the ceremony, the officers read a portion of my application that highlighted my classroom's vertical indoor hydroponic system and donations. I used broken string instruments to study sound waves, giant sloth fossils to study evolution, and two dozen rubber ducks to study genetics. Other donations included a bearded dragon and eighty pounds of art supplies.

To reach different learning modalities, I encouraged students to draw thirty-foot DNA strands on the sidewalk, create raps/songs on ecology, and draw on the windows to show incomplete and codominant traits. Students' art projects, such as a puppet show of Florida animals and ecology, were sent to students in Chicago, New York, and Turkey. Those students responded with what they were learning, creating a meaningful experience for my students. I broke ground to install four sixteen by four foot vegetable and herb gardens.

To supplement a teacher's pay, I was continuing to write for a half-dozen magazines. My students were shocked that I interviewed a veterinarian living in Massachusetts who amputated a broken leg of a chicken and later attached a leg generated from a 3-D printer. This further showed my students the application of science in the real world.

Bertha Vázquez, director of the Teacher Institute for Evolutionary Science, introduced herself after the FAST award ceremony and congratulated me. She told me that I was doing great work and that we need more teachers like me in the world.

One year later, I attended the FAST conference again, this time accepting a classroom mini grant. I reconnected with Bertha and offered my skills to improve TIES' social media. A few weeks later she hired me as *the* TIES associate. My duties included expanding TIES workshops to all fifty states. With my zoological connections, I was able to help secure states including Alabama, Massachusetts, Montana, and Utah.

In my school district, I helped create a six-hour Darwin Day celebration which combined the local university, museum, national guest speakers and TIES presentations and labs. The school district paid the teachers and provided lunch and continuing education credits. Bertha and I presented the first year and since then I have been instructing my peers. One teacher, so impressed with the TIES curriculum we presented, said, "My school will want you." A few months later, I was offered the job.

Introducing Evolution

 I, like Bertha, enjoy starting the unit of evolution with the flea bell-ringer. Since it apparently only talks about pets and fleas, the unit is a relatively gentle way of introducing what is often a controversial subject. Fleas that happen to have a mutation that gives them resistance to a chemical is an idea that students can understand. Students most likely witnessed it themselves. Most children can understand that organisms have changed over time. What is harder for students to grasp is that hominids have also changed over time. Because of religious beliefs or gut feelings they may think that humans are immutable. Through the flea bell-ringer and after the introduction of evolution, they'll understand the concept of natural selection applies to all organisms.

Many believe it would be best practice to sprinkle the study of evolution throughout the year, rather than clump it into one unit, which could potentially be skipped by a fearful untrained teacher or squeezed in at the end of the year because of state testing and other pacing guide hurdles. In addition to life science, I teach separate classes of agriculture. I find it quite easy to oscillate between evolution and agriculture. In the TIES Evolution unit, there is one slide on artificial selection. While I do not show them this slide until I am delving into the evolution unit, I do verbally reference the examples often. The slide shows a hieroglyphic of Egyptians milking cows. Humans domesticated cows around ten thousand years ago and started milking them between six and eight thousand years ago. Back then, cows were producing a little bit more milk than the calves needed to survive. The average cow today in the US produces 2,500 gallons a year! The daily average is eight gallons or seventy pounds. After, but sometimes before, students learn genetics, I ask, "How did we get cows today that produce so much milk." Through discussion, several students in each class can answer something along the lines of selective breeding, eat the ones that don't produce extra milk, and keep and breed the heavy milkers.

The artificial selection slide also has rainbow display of carrots and photos of wild mustard relatives. While working in the school garden, I'll occasionally mention that carrots started off being tiny discrete roots that where purple. Even yellow and white roots came before orange. Asking students, "How did farmers get thick juicy orange carrots from tiny purple roots?" will get them imagining all of the incremental steps growers took to breed a carrot. This sets them up to better understand evolution. I take this impromptu carrot lesson to also teach them about flower reproduction, biennials, and the function of roots, leaves, and stems. Teacher tip: Any lesson that involves eating the subject is a winner.

In addition to the agriculture students, science students also visit the fifteen raised vegetable and herb beds weekly during class. After seeing snapdragons exclusively referenced in the genetics section of our textbook, we started growing several varieties. We also grow broccoli, cauliflower, cabbage, kale, kohlrabi, and brussels sprouts. All these leafy greens are descendants of a wild mustard plant. I tell students that distinct cultures and countries in the Mediterranean took the wild mustard plant and bred them to serve unique purposes in their kitchens. Seeing how similar the seeds, seedlings, flowers, and leaf structures are reinforces the plants' relatedness. It does not take much prodding to get students to discuss how a farmer, or backyard gardener, could select for different favorable traits, and create something that looks and tastes drastically different than their neighbor's crops.

Favorite Investigative Unit or Activity

If I only had time for one activity in the evolution unit, I would choose the rotation lab featured on the TIES website. The other life science teacher at my school, Kim Marr, helped me create this

VIDEO RECOMMENDATIONS

I wrote a few TED-Ed lessons in the rotation station activity. One of my favorites is titled, "The Wild World of Carnivorous Plants." While it speaks to great depths of adaptation in the plant world, it only touches the surface of carnivorous plant evolution. You can freeze an accurate cladogram to highlight convergent evolution and analogous structures.

The real reason I am mentioning this TED-Ed lesson is to highlight the ed.ted.com website. There you'll find all your favorite YouTube TED-Ed lessons, but you'll also find five multiple choice questions, two short-answer questions, and a long-response question for each video to use in your classroom.

resource. We compiled things that she had been using for a long time and I included some singletons from the TIES website. Kim and I created a two-three-day, student-driven station rotation lesson. These stations complement the TIES unit but we usually have the students do the activities before exposing them to the direct teaching. By having the students participate in the rotations, everyone has something to pull from and contribute as we go through the slideshow. Each station takes about fifteen minutes. We plan on two to three a day with a debrief at the end of the class. We have made it an annual tradition to bring both of our classes (forty to fifty students) to the media center to work on the eight stations simultaneously. Slides 2-8 in the rotation lab should be laminated and placed at stations throughout the classroom. Each student should receive slides 9-16 to serve as their worksheets. Answers and instructions are written in the comment section below each slide. We found it impactful to have the media center specialist collate their worksheets into a spiral-bound notebook at the end of this section and have students place a cover on the front that reads, "My Evidence for Evolution."

Dealing with Conflicts

One of the biggest hurdles to teaching evolution is the preconceived feelings that students bring into the classroom. At open house before school starts I have had parents tell me that they are one type of religion and when I get to evolution, just know that their child will not be believing anything I say.

The main ideas that students must learn: evolution is supported by multiple forms of evidence; natural selection is a primary mechanism leading to change over time; and the scientific theory of evolution is the organizing

The TIES unit is a thorough compilation of resources. Break up the unit with the hyperlinked videos, labs, and activities to keep engagement high.

BOOK RECOMMENDATIONS

I presented a day-long TIES workshop at Utah's Hogle Zoo. While in Utah, I visited the area's attractions and stopped into the local bookstore. There I saw a mesmerizing display of a newly released book, *Unnatural Selection (2018)* by Katrina van Grouw. The book is a beautiful blend of art and science. I would consider it a coffee table book, but it would be well suited to any classroom. The book celebrates the 150th anniversary of Charles Darwin's work, *The Variation of Animals and Plants under Domestication* (1868). I was intrigued by the book's artwork, and I proposed to Bertha that we contact the author and add her to the queue of our monthly webinars. Katrina was wonderful and obliging. As a nonartist, I could not even begin to calculate how Katrina not only rearticulated the skeletons but sketched them so wonderfully. The book includes more than four hundred illustrations of living animals, skeletons, and historical specimens. Many including the Texas longhorn cattle, were assembled in Katrina's home! The book is not only about pets and livestock. The book details how humans can not only change animal's behavior and coat color, but also their bone shape, size, and quantity. This is the perfect book for a fellow life science teacher because it can be used in the classroom to show what humans are capable of in a few short centuries. Think of the over 200 dog, 350 pigeon, 450 chicken, or 1,000 cattle breeds, and now look around and see what nature has done in a few billion years.

principle of life science. This last benchmark allows me to easily weave evolution into other concepts to build a stronger foundation. One reason evolution is difficult to teach is that it is called the "theory of evolution." In everyday language, a theory might mean a guess or hunch. But in science, a theory refers to a well-supported explanation and says why something happens, explains nature, and is evidence-based.

Another misconception is that humans descended from chimps, which immediately puts the students in a defensive stance, unwilling to compromise or learn. Evolution states that all life on Earth share a common ancestry. I show my students evidence such as DNA, fossils, and cladograms to show my students we have not descended from monkeys but rather are distant cousins.

Another hurdle is that my students often have trouble with the nature of science, timelines, interpreting graphs, reading charts, etc. They memorize dates and figures, but they do not truly understand concepts such as the vast scope of Earth's or humankind's timeline. They find it hard to understand that I did not have a flat-screen TV or cell phone growing up, let alone that Earth is 4.5 billion years old. Interactive timelines, videos, and in-class models help with this obstacle.

Evolution *ties* the life sciences together and is a seamless underlying thread. It should be taught throughout the year and not as a singleton. Evolution *ties* into genes and heredity. Once students understand diversity and mutations, they can see how artificial selection and natural selection work. Evolution also easily relates to benchmarks concerning Earth's timeline, fossil formation, and the rock unit. When I teach the law of superposition and radiometric dating in the fall, I tell students that these concepts will be popping up again in the spring with the evolution unit so pay attention!

I have a gap in my own teaching practice since I became a certified middle school science teacher by earning a BS in animal behavior and passing a science certification exam. I did not focus on many of the subjects that I am now required to teach. TIES has provided me the resources to feel confident in teaching evolution. TIES has also provided leadership opportunities which have led to a new teaching job and a science education foundation fellowship.

5 Teaching Darwin's Theories

Robert A. Cooper, MS in Biology
Warminster, PA

One Sunday evening in December 2016, as I lay drifting off to sleep, I was suddenly jolted out of bed by sharp pains in my lower back, right hip, and running down my right leg. Mild bouts of pain in these areas were no stranger to me. They had occurred chronically over the previous ten or twelve years, but were never as severe as this. I spent the rest of that December night standing at the kitchen sink trying, without success, to find a comfortable position that would relieve the agonizing pain. Early the next morning, my wife took me to the emergency room where the doctor informed me that I had herniated disks in my lower back.

Now that time and surgery have put some distance between the present and that painful experience, I can reflect on the cause of my back problems. The most immediate cause, of course, was pressure on the sciatic nerve caused by bulging cartilage disks between my vertebrae. The disks normally act as shock absorbers, but injury or aging can cause the disks to slip, bulge, or rupture resulting in pressure on the nerves that emerge from the spinal column and pass through narrow passages between the vertebrae. This causes pain, numbness, and tingling. Low back pain is a very common malady. About 80 percent of adults have low back pain at some time in their lives (National Institute of Neurological Disorders and Stroke 2018). But why should low back pain be so common? And why was I afflicted with this pain? I had not done anything strenuous, nor had I injured my back in any way. But age was working against me. My spine had been supporting my body for fifty-nine years, and those years of pressure on the spongy disks that separate the vertebrae caused two of them to bulge and press on nerves.

Why aren't our bodies built better to withstand the pressure our upright posture puts on these disks so they last longer? (Olshansky, Carnes, and Butler 2001). For that matter, why do people get heart disease, diabetes, or cancer? In many respects, the human body appears to be a marvel of design. Your heart beats about one hundred thousand times a day and about thirty-five million times in your lifetime, pumping the blood that transports nutrients and oxygen to your tissues and carrying away metabolic waste. If you've ever seen a patient attached to a heart monitor, you've seen how the slightest body movements cause the heart rate to increase, making adjustments to ensure an adequate supply of oxygen and nutrients to the tissues at all times. Considering the many marvelous features of the human body and contrasting them with the existence of disease prompted physician Randolph Nesse and evolutionary biologist George Williams to ask, "Why, in a body of such exquisite design, are there a thousand flaws and frailties that make us vulnerable to disease?" (Nesse and Williams 1994, 3).

To answer their question with regard to my back problems, we have to adopt an evolutionary perspective. Our bodies are products of evolution, not rational design. Our earliest vertebrate ancestors were fish,

BOOK RECOMMENDATIONS

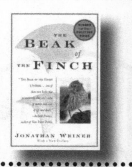

For natural selection: an excellent book providing detailed background on Peter and Rosemary Grant's work with Darwin's finches is Jonathan Weiner's *The Beak of the Finch: A Story of Evolution in Our Time* (1994).

For dinosaur-bird transition: *The Rise and the Fall of the Dinosaurs* by paleontologist Steve Brusatte.

whose spine was not subjected to the compression forces our spines must endure (Shubin, 2008). Terrestrial vertebrates inherited their spines from fish, with modifications as species adapted to the conditions found on land, but most of them were quadrupedal and, like fish, their spines were not subjected to the compression forces that ours are. It was our australopithecine ancestors who evolved bidepalism. When they began to walk on two legs, their spines could not be completely refashioned to produce an optimal design for upright walking (Olshansky, Carnes, and Butler 2001).

But issues with our spine that lead to back problems are not the only glitches with which we must contend. Both our bodies and minds are products of the historical process of evolution, and their "design" reflects many constraints and compromises resulting from our evolutionary ancestry (Shubin 2008; Marcus 2008; Lents 2018). Selection only favors modifications to existing structures that are sufficient to allow survival and reproductive success. In other words, evolution tinkers with existing structures to produce satisfactory solutions to problems; it does not redesign from scratch to produce optimal solutions (Jacob, 1977). The prevalence of many human glitches and ailments makes more sense when we consider them from an evolutionary perspective (Nesse and Williams 1994).

Introducing Evolution

I focus first on natural selection and then consider descent with modification in a later section. Although natural selection seems a simple idea to those who understand it and accept its implications, intuitive concepts and modes of reasoning can make it very difficult for novice learners to develop an accurate understanding of the process

(Shtulman 2017). Understanding natural selection requires that students think in ways that are different from the way in which they normally make sense of the world. Children have a natural tendency to explain events in terms of intentions or goals of a central actor, and this tendency can persist into adulthood (Bloom and Weisberg 2007; Sinatra, Brem, and Evans 2008). The application of this intuitive mode of reasoning may cause students to develop misconceptions about processes such as natural selection. One common student misconception is that environmental pressures cause a need for change and all individuals in the population simultaneously respond to this need by adapting, i.e., modifying their features, to survive. This statement reflects goal-oriented thinking and it stems from the fact that students' intuitive theories cause them to focus their attention only on the individual organisms, when they should be dividing their attention between the fate of the individual organisms and the resulting changes in the makeup of the population (Cooper 2017; Lucci and Cooper 2019).

I would choose an introductory activity that enables students to confront this common misconception and begin to develop a more accurate conception of the process. I would have my students observe, analyze, and discuss the histograms shown in figure 1. The histograms illustrate the evolution of beak depth in the medium ground finch (*Geospiza fortis*) under drought conditions on the Galápagos Islands. Guided inquiry instruction in interpreting these histograms will help students develop an understanding of natural selection consistent with what is called for in performance expectations MS-LS4-4 and MS-LS4-6 in the Next Generation Science Standards (NGSS Lead States 2013; See Appendix).

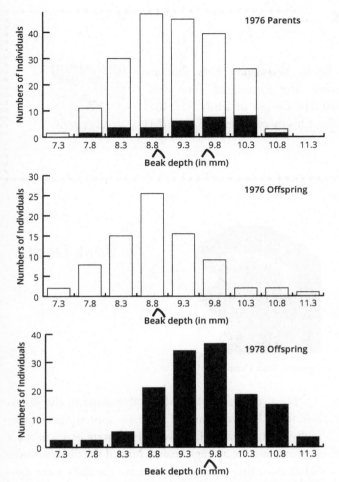

Figure 1: Evolutionary change in beak depth in the population of medium ground finches (*Geospiza fortis*) on the island of Daphne Major (Grant and Grant 2003).

The following story of the 1977 drought and its effect on the medium ground finches serves as an anchoring phenomenon for investigating natural selection. Begin by having students read the following passage:

Not long ago, if you developed a bacterial infection you could take an antibiotic and know, with confidence, that the medication would cure the infection. But in recent years, many of the bacteria that cause disease have become resistant to antibiotics. Similarly, bed bugs have become resistant to insecticides, and tumors have become resistant to the drugs doctors

use to treat cancer. Why is this happening? To understand these issues, it will help to investigate a phenomenon observed in some birds that live on the Galápagos Islands.

In 1973, Peter and Rosemary Grant, two scientists from Princeton University, began a scientific investigation of the birds known as Darwin's finches. Darwin's finches live on the Galápagos Archipelago in the Pacific Ocean six hundred miles west of Ecuador. The Grants chose to work on the island of Daphne Major because of its small size and isolation from human activity. They reasoned that the small size of the island would enable them to capture, band, and measure every individual bird on the island. They measured each bird's weight, wing length, leg length, beak length, beak width, and beak depth. For the first four years, they observed very little change. Then, in 1977 a severe drought that lasted eighteen months struck the island. The Grants observed significant changes in the numbers of individuals of all species living on the island. But most importantly, they documented significant changes in the population of the birds known as medium ground finches (*Geospiza fortis*).

Medium ground finches are seed eaters. Daphne Major is normally populated with a variety of cacti, grasses, and shrubs that all produce seeds that the birds eat. The seeds produced by the various plants have a wide range of sizes and hardness, and the seeds support a population of medium ground finches with a range of beak sizes. But the drought caused a change in the vegetation and a resulting change in the types of seeds available for the finches to eat. As the drought wore on for eighteen months, all of the grasses and shrubs died, leaving only the cacti, and the range of seeds available to eat was more limited. Toward the end of the drought there were only large, hard seeds of the cactus plants left. The scarcity of food, and the fact that there were only very large, hard seeds posed a challenge for the medium ground finches.

VIDEO RECOMMENDATIONS

Three excellent short films related to Shubin's book, *Your Inner Fish,* that are suitable for the classroom are "Great transitions: The Origin of Tetrapods" (HHMI BioInteractive, 2014b), "Great transitions: The Origin of Birds" (HHMI BioInteractive, 2015b), and "Great transitions: The Origin of Humans" (HHMI BioInteractive, 2014a).

After students have finished reading, have them answer the following two questions:

1. What will happen to the finches with smaller beaks now that there are only large hard seeds to eat? Will they be able to adapt? If so, how will they adapt? Make a prediction and write it in the space below.
2. What other questions do you have about the drought on Daphne Major and its effect on the medium ground finches?

After students respond to the questions, take time to discuss their predictions and any questions they may have about the story. Ask some students to share their prediction with the class. Student predictions can be written on the board, without evaluating them, and revisited in a discussion at the end of the activity. Then provide each student with a copy of the histograms in Figure 1 and explain that they were produced using data collected by Peter and Rosemary Grant. These data will help them test their predictions.

A brief orientation to the histograms will help students get their bearings. Begin by discussing the axes of the histograms to ensure that students understand the variables displayed on each axis. The x axis of each histogram records beak depth, the distance from the top of the beak to the bottom at its greatest extent (See Figure 2). This trait was found to be the most significant of the many traits investigated by the Grants. The y axis of each histogram records the number of birds at each beak depth measured in millimeters (mm).

Figure 2: Beak Depth

Next, discuss with students the shape of the distributions of beak depths. All three are approximately normally distributed (bell curves). The carets beneath the x axes indicate the average beak depths of the distributions. What does this tell you about how the data were generated? Data that follow a normal distribution are generated by a random process. In this case, the normally distributed beak depths are generated by the randomizing processes of meiosis and sexual reproduction. The range of beak depths in the finch population shows the variation in this trait. You can help students understand this by calling their attention to variation in the heights of humans and showing them a histogram of human heights. Remind them that children tend to have heights similar to their parents, and the same is true for the beak depths of the finches. But, in each case there is considerable variation in the population.

Finally, inform students that the white bars in the histogram showing the 1976 parents (top panel) represents the number of birds at each beak depth before the drought occurred, and the black bars show the number of birds remaining at each beak depth after the drought that began in January of 1977 and lasted 18 months. The

1976 parents had an average beak depth of approximately 8.8 mm, and the birds remaining after the drought had an average beak depth of approximately 9.8 mm. The histogram in the middle panel (1976 offspring) displays the distribution of beak depths of the offspring of the 1976 parents produced before the drought, and the histogram in the bottom panel (1978 offspring) shows the distribution of beak depths in the offspring of the 1976 parents that survived the drought. Have students work in small groups of two to three students each to answer the following questions.

Questions for small group discussion:

Q1: There is a considerable difference in the number of birds represented by the white bars and the black bars on the histogram. What happened to the birds from 1976 (white bars) that are not shown in 1977 counts (black bars)?

Model Response: Most of the birds on the island in 1976 died as a result of the drought, hence the significantly lower numbers of survivors in 1977 following the drought.

Note: Some students may suggest that many of the birds migrated elsewhere to find food. This is unlikely. The distance between Daphne Major and the nearest islands in the archipelago is great enough to make migration very challenging, and birds undernourished because of the drought would not be able to make the trip. Even if they could, the weather conditions causing the drought impacted the entire archipelago, and the archipelago is six hundred miles from the coast of Ecuador, so they could not fly to South America for food.

Q2: Could the birds grow larger beaks to survive the drought? If they did, would you expect the pre-drought (white bars) and post-drought (black bars) histograms to look as they do? If they were not able to grow larger beaks, how would the post-drought distribution of beak depths compare to the pre-drought distribution?

Model Response: Individual birds cannot grow larger beaks to crack large, hard seeds any more than a slow runner could suddenly develop Olympic-class speed to fulfill a dream of becoming a wide receiver on a professional football team. If all of the birds could "adapt" in this way, then why would any of them die? If individuals could modify their beaks, the number of birds would be the same before and after the drought. All the birds would adapt and survive, and the only change would be that the entire distribution would shift to the right toward a higher average beak depth. The data does not support this claim. Most of the birds died during the drought.

If the birds are not able to grow larger beaks, the histograms would look exactly as they do. Most birds died, but there were differences in terms of the ability to survive during the drought. On the average, birds with larger beaks tended to survive at higher rates than birds with smaller beaks. The group with the highest survival rate had beak depths close to 10.3 mm Approximately one-third of the birds in this class survived, while all other classes had much lower rates of survival. The survivors reproduced and, as a result, the distribution of beak depths shifted toward a higher average beak depth.

Q3: Is there evidence that beak depth is an inherited trait? Why is there a difference in the average beak depth of the offspring produced in 1976 and 1978?

Model Response: The parents of the offspring produced in 1976 are represented by the white bars in the top panel of figure 1. The parents of the offspring produced in 1978 are survivors of the drought and are represented by the black bars in the top panel of figure 1. In each case, the similarity in the distributions of parents and offspring (with the same average beak depth) suggests that beak depth is inherited. The difference in average beak depth of the offspring produced in 1976 and 1978 was caused by the drought. Birds with smaller beaks died at a higher rate than birds with larger beaks. The survivors, with a larger average beak depth, produced a new generation with a similarly larger average beak depth.

Q4: Write a clear and thorough explanation, consistent with the data in figure 1, of how natural selection caused the differences in the distributions of the 1976 parents and the 1977 survivors, and the differences between the 1976 offspring and the 1978 offspring. Be sure to make reference to the data in your response.

Note: To scaffold students' efforts to write good explanations for cases where natural selection occurs, I encourage them to structure their explanations using the mnemonic device VISTA (American Museum of Natural History, 2005). This will ensure that they include all of the essential elements of a good explanation. VISTA stands for variation, inheritance, selection, time, and adaptation. This mnemonic captures most of the essential elements of Darwin's theory of natural selection. Applied to the finches during the drought, the mnemonic would produce an explanation that looks something like this:

Model Response: All three of the histograms are approximately normally distributed and display a range of **VARIATION** (V) in beak depth from 7.3 mm to 10.8 mm in the case of the 1976 parents, and a range of 7.3 to 11.8 mm in the cases of the 1976 and 1978 offspring. The distribution of the 1976 offspring resembles that of the 1976 parents and they have the same average. In addition, the distribution of the 1978 offspring resembles that of the 1977 drought survivors and they have the same average. These similarities suggest that beak depth is an **INHERITED** (I) trait. The drought imposed strong **SELECTION** (S) on the finches and favored the birds with greater beak depths. A majority of the birds died during the drought, and birds at the lower end of the distribution, with smaller beaks, died at higher rates than those with larger beak depths. Selection caused the average beak depth of the population to shift from approximately 8.8 mm before the drought to approximately 9.8 mm after. This evolutionary change in beak depth took one generation. Evolution by natural selection involves changes in the average traits of a population over **TIME** (T) (generations). Individuals cannot evolve during their lifetimes. As a result of the change in average beak depth, the population was better **ADAPTED** (A) to the prevailing environmental conditions immediately following the drought.

Favorite Investigative Unit or Activity

If I had time enough for just one activity to introduce students to descent with modification and the tree of life, I would introduce

these concepts with an activity called "What Did *T. rex* Taste Like?" from the University of California Museum of Paleontology (https://bit.ly/trex-tastelike). The activity introduces students to cladistics, a method used to determine evolutionary relationships between groups of organisms, and classify them based on shared inherited features. In cladistic analysis, biologists use molecular, biochemical, physiological, anatomical, or behavioral features to compare groups of organisms and generate hypothetical evolutionary trees illustrating how the groups are related by evolutionary descent (See appendix for the NGSS Disciplinary Core Idea and the related Performance Expectation MS-LS4-2). Groups of organisms that share a set of features and all derive from a single common ancestor are referred to as clades. Students who complete the *T. rex* activity will gain a better understanding of how to correctly interpret evolutionary trees, or cladograms, and also understand the value of evolutionary trees to biologists.

"What did *T. rex* Taste Like" begins by introducing students to the vast diversity of living things and explains that all species can be traced back through lineages, lines of descent, to a single common ancestor. A comparison is drawn between family trees, or pedigrees, and evolutionary trees. Two children (siblings) from the same family resemble one another because they inherit some common features from their parents, but they also differ from each other and from their parents. Similarly, two closely related species, like coyotes (*Canis latrans*) and wolves (*Canis lupus*), resemble one another because they have inherited some common features from their most recent common ancestor, but they also differ from one another and from their common ancestor. Coyotes and wolves are sibling species. This claim about coyotes and wolves along with its supporting evidence provides a specific example of the general argument that Darwin (1859) makes in the *Origin*, his *One Long Argument* (Mayr 1991). There is no mysterious, supernatural principle guiding evolutionary change. It is simply an extension of the observable facts of biological reproduction and variation extended over long periods of time. In other words, descent with modification.

The main part of the *T. rex* activity introduces students to basic terminology and explains how to correctly read evolutionary trees. Finally, as a culminating special assignment, students are challenged to determine which group in a vertebrate evolutionary tree is most closely related to *T. rex*. As they complete the activity, students learn that birds and theropod dinosaurs, such as *T. rex* and *Velociraptor*, share the most features in common and are therefore close cousins. In fact, birds are considered avian dinosaurs (University of California Museum of Paleontology 2020). Similar to our understanding of the relationship between coyotes and wolves, the logic supporting the claim that birds and dinosaurs are closely related rests upon our understanding of biological reproduction, variation, descent with modification, and the evidence from the fossil record, which demonstrates the anatomical similarities between birds and theropod dinosaurs. Although, a close relationship between birds and theropod dinosaurs was first proposed by Thomas Huxley in the 1860s, it remained controversial until a remarkable collection of fossils was found in Liaoning Province, China, in the mid-1990s. These fossils provide strong evidence supporting the bird-dinosaur relationship. In his book, *The Rise and the Fall of the Dinosaurs,* paleontologist Steve Brusatte explains,

"The Liaoning fossils sealed the deal by verifying how many features are shared uniquely by birds and other theropods: not just feathers, but also wishbones, three fingered hands that fold against the body, and hundreds of other aspects of the skeleton. There are no other groups of animals—living or extinct—that share these things with birds or theropods: this must mean that birds came from theropods. Any other conclusion requires a whole lot of special pleading" (Brusatte 2018, 282).

Although it's possible that the many features uniquely shared by birds and dinosaurs, including complex structures such as feathers, evolved independently more than once, it's not very likely. The simplest explanation is that birds and dinosaurs inherited these shared features from a single common ancestor. In other words, birds and theropod dinosaurs belong in the same clade. Birds are dinosaurs.

After using evidence to determine the relationship between birds and dinosaurs, the *T. rex* activity illustrates for students the value of evolutionary trees to biologists by having them answer questions about soft tissue features, physiology, vision, behavior, and feathers on *T. rex*.

No one has ever seen a living *T. rex*, so these questions cannot be answered by observing a *T. rex* directly. All that we have are fossils. But, knowledge of the fact that birds and dinosaurs such as *T. rex* are close relatives enables students to propose likely answers to questions like, "Did *T. rex* have an amniotic egg?"; "Was *T. rex* warm-blooded or cold-blooded?"; "Could *T. rex* have had feathers?"; "Did *T. rex* have color vision?"; "How many chambers were there in *T. rex*'s heart?"; "Did *T. rex* sing to its offspring?" These questions are provided in the teacher's guide as an assessment activity.

After completing the *T. rex* activity, students should watch the short film *The Origin of Species: The Beak of the Finch.* This film serves as an excellent culminating activity that ties together the students' guided analysis of the histograms and the *T. rex* activity. Biologist Sean Carroll, who narrates the film, asks, "So how does one species split into two? A typical scenario is that two populations become separated geographically, and undergo enough change in their respective habitats, that if or when they come into contact again, they do not mate" (HHMI BioInteractive, 2013, time stamp 11:25). Extended over eons, this fundamental process of speciation is responsible for the vast diversity of life we find on Earth.

A fundamental concept that students must understand as an outcome of engaging with the *T. rex* activity is that one does not have to be a time traveler to know that birds originated from the theropod dinosaurs. Scientific knowledge of the past is possible because past events leave behind trace evidence indicating that they occurred (Cooper, 2002, 2004). The ability of scientists to know what happened in the past depends on their ability to find and correctly interpret the trace evidence. Clearly, fossils provide trace evidence that the critters known as theropod dinosaurs once lived on Earth. But just as birds share anatomical features with dinosaurs that show their common ancestry, every living critter currently on Earth also carries trace evidence of its origin, and its kinship with other living things, in its biochemistry, physiology, anatomy, and behavior. The methods of cladistics enable scientists to analyze and interpret that trace evidence. Shared features derived from a common ancestor are called homologies. Living things share homologous features because of the DNA they inherited from ancestors stretching back into the deep recesses of Earth's history. This is why Carroll (2006) referred to DNA as "the ultimate forensic record of evolution."

Dealing with Misconceptions About Evolutionary Trees

Students' tendency toward goal-oriented thinking also influences their pre-instructional understanding of descent with modification. Rather than viewing evolution as a tree, students see it as a linear process where the goal is to transform primitive ancestral species into more highly evolved forms. Evolution is represented as a ladder of progress, with humans as the inevitable and superior product of the process (Gould, 1989). In images, this view of evolution is typically represented by a horizontal series of primates beginning with a chimpanzee on the left, which is transformed into a series of more human-like ancestors, and finally to a contemporary human. This view is reinforced by a majority of the images students regularly encounter in the wider culture where attempts are made to portray evolution. Images like this are no doubt partly responsible for the recurring question from anti-evolutionists, "If we came from chimpanzees, then why are the chimpanzees still here?" Posing this question for students to consider is a good way to introduce a lesson on descent with modification, and bring students misconceptions forward for examination. Richard Dawkins provides a clear response to the question in a brief video: *Why are There Still Chimpanzees?*. Having students watch this video as one of the culminating activities of a lesson on descent with modification is an effective way to counter one of the most pervasive misconceptions about evolution.

Introducing students to evolutionary trees using the *T. rex* activity is a major step toward improving their understanding of evolution. However, in addition, it is also necessary to directly confront their misconceptions.

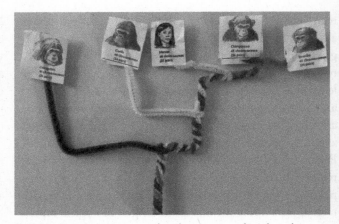

Figure 3: A student-constructed evolutionary tree based on the activity described in Halverson (2010).

For example, looking at the evolutionary tree of primates shown in figure 3, students may have the tendency to read left to right across the tips of the branches and maintain the view that evolution is a progressive process leading inevitably toward humans. You can help students confront and dispel this misconception by having them construct their own evolutionary tree similar to the image above using pipe cleaners (Halverson, 2010). With the pipe cleaner tree, students can see that the branches can be rotated around any node and the order of the groups from left to right across the tips is meaningless. All that matters is the branching order, and that is not changed by rotating the nodes.

Conclusion

Developing an accurate understanding of evolution can be challenging for students. Evolutionary concepts often seem counterintuitive. One of the most difficult misconceptions to overcome is the idea that evolution is a goal-oriented process leading progressively to humans. The activities described here can help students confront and overcome misconceptions related to goal-oriented thinking about natural selection and descent with modification. Students just need a clear demonstration of how their initial understanding fails to adequately explain the phenomenon in question, followed by a clear demonstration of how the scientific theory adequately accounts for the phenomenon.

6

An Ode to the Unbroken Thread

Chance Duncan, MS in Science Education

Arkansas

To me, being a science teacher isn't a job in the traditional sense of the word. A job strikes me as something a person does to make a living, whether or not the person actually enjoys it or regardless of how important it is. Teaching science, however, is a way to love living. I could wax on philosophically and socio-politically about how important a good, proper science education is, but I'll leave it to one of my personal heroes, Neil deGrasse Tyson, who said it best: Science literacy is a vaccine against the charlatans of the world that would exploit your ignorance. That just about sums up the importance of my job to me and why I have devoted my life to this endeavor. Though I teach biology, I always have to ask myself if it really is that important for my students to understand the intricacies of the physiological processes involved in combusting glucose to generate ATP? I mean, personally, yes, I think that's important to know, but is that really important for tenth graders to know? Or is it better that they learn about those processes in the context of developing a better sense of awe and wonder about the nature of reality? I think the latter is the most important reason we teach science.

Science has played a very important role in my life for as long as I can remember. I had the fortune to grow up in rural Arkansas. My backyard consisted of acres of fields and woods. Wildlife captured my attention from my earliest memories, and I spent as much time as I could in nature. Little did I know it, but I was laying the groundwork for what would become a lifetime of wanting to help others understand and appreciate nature.

I was one of those kids who was happier to watch a documentary on the Discovery Channel or Animal Planet than to watch cartoons. Of course, that was back when those channels focused on real science, not finding Bigfoot or evidence of ancient astronauts.

These documentaries are how I was first introduced to the concept of biological evolution. I count myself lucky that I was not raised in a religious environment, so I never saw an issue with the scientific understandings portrayed in the videos. To me, it was patently obvious that evolution was real and was simply how organisms evolved to deal with changing environments. I recall wanting to do a science fair project dealing with human evolution when I was in the fifth grade, and naively failing to understand why my teachers were uncomfortable with it and persuaded me to do something rather pointless with a solar oven.

My middle school years were when I discovered that not all people accepted valid science. I started hearing rumors about a teacher I would have in a couple of years who was an "atheist" (he wasn't) and who dared not to let students write "God did it" on questions about how the earth was formed (he didn't). It slowly dawned on me that many of the understandings I simply took for granted were actually wildly controversial to some of my peers; and alas, to some of my teachers.

BOOK RECOMMENDATIONS

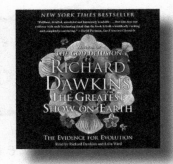

There is a wealth of books dealing with the science of evolution, from exploring the history of the discovery of the idea by Darwin, Wallace, and others, to probing the various lines of evidence and showing how we came to the conclusion that life is not immutable after all. My favorite would have to be Richard Dawkins' *The Greatest Show on Earth* (2009). I have long been a fan of Dawkins' work. While some find his particular method of explanation and illustration to be off-putting, I very much appreciate his unapologetic approach to explaining what we know about the natural world. He has a way of using metaphor to help readers understand some of the more difficult aspects of evolution, some that teachers need to be ready to help students understand. He walks the reader through a thought experiment where he describes a rabbit and her progeny, all of whom are rabbits, and her parents, both of whom were also rabbits. But, as he says, if you stretch forward a thousand generations or reverse a thousand generations, what you would see would no longer be a rabbit. In fact, if you go back far enough along this line of descent, there would be an alternate path you could take that would end in a leopard! I love taking students through this thought experiment when we're discussing speciation. It is so difficult for them to imagine how humans could share common ancestry with other primates, and I almost invariably am asked the "Then why are there still monkeys?" question. In a way, I understand their frustration. The idea that my parents were human and my children would be human, yet at some point in either direction they wouldn't be, is difficult to comprehend! Examples like those that Dawkins provides help them develop a better grasp on how small changes, amplified over enormous amounts of time, can result in profound diversity.

Throughout middle and high school, I never really learned about evolution in classrooms. I taught myself through what was then a fairly young internet, and I engaged in countless debates with my peers about why evolution was so obviously real. These debates ended some friendships but generated others. My best friend throughout high school started our relationship through an effort to convert me to her religion. She brought her youth pastor to my house to try to share with me the "Good News." She made a valiant effort, but in the end, it backfired somewhat spectacularly. By the end of our high school tenure, she seemed to be leaning toward agnostic, and shortly thereafter, lost her faith completely.

It wasn't until my undergraduate biology courses in college that I finally had teachers who were willing to broach the topic of evolution, meaning I was woefully underprepared for these kinds of courses. I struggled through them, obtained my degree in science education, and am now in my eleventh year in the public school classroom attempting to inspire my students to recognize and respect the amazing nature of reality.

I could not possibly imagine how a teacher could educate a student about biology without the unifying, underlying theme of evolution. It ties concepts together in a way that is both simple and wildly complicated. Without this theme, biology would just be a list of random disconnected facts for students to memorize. I fear that this, which was the case for me in my own high school education, is still the case for many students around this country. I can confidently say, however, it will never be the case for my own.

My motto: Say it early, and say it often! The "it" in this instance is evolution. Every single thing I teach in

VIDEO RECOMMENDATIONS

Much like the namesake of this chapter, my favorite short video on evolution would have to be John Boswell's "The Unbroken Thread," the fourth in his Symphony of Science video series. This series gives viewers a glimpse at the awe and wonder of how science has and continues to improve our lives. He does this by combining clips of notable scientists and science communicators played over spectacular imagery. He creates a melody and auto-tunes the voices to match, from which results a very cool sciencey music video.

I use these videos each year to introduce or enhance topics ranging from the nature of science to atomic structure, evolution, and cosmology. One year I even had students designing backgrounds for their personal laptops with the Dawkins' quote "Science is the poetry of reality" from John's video by the same name. I believe John's work does an excellent job capturing the excitement that science can bring to those who choose to use it as a tool to investigate the world.

biology is through the lens of evolution. By the time we get to the evolution unit, students have heard the word so often that it loses much of the negative connotation they've been taught. Even the religious fundamentalists I have in class will usually start to use phrases like "they evolved to" when answering questions. I have known teachers who refuse to say the word "evolution," and instead use euphemisms such as "natural selection," "adaptation", or "change over time." I really don't think they're doing themselves any favors. Shying away from the word is belittling the most important concept in all biological science. Students should be using evolutionary verbiage to explain their understanding of every concept in biology, from cellular structure and energetics to ecological interactions. This consistent use will help the teachers make an easier transition to the evolution unit, and will help the students by giving them a unifying concept around which they can connect everything they learn in class.

Introducing Evolution

I teach high school biology at a medium-sized school district in central Arkansas. Most of our biology students are in the tenth grade. Though the curriculum we use has a unit dedicated to evolution, it is definitely not the first time my students are introduced to the topic. In fact, when students come

downstairs to our STEM Center, they are greeted by a display of hominin skull replicas.

My colleague Mark Meredith (owner of biologyalive.com) and I spent a couple of years putting together the funds to buy a set of hominin skull replicas from Bone Clones Inc. We created the display so that students and visitors could walk by and observe artifacts of our evolutionary history daily. We also developed a lab activity for the evolution unit later in the year we call the "Family Reunion" (appendix 1) in which students measure morphological features such as facial prognathism, dental arcade arch, and forehead slope to hypothesize a potential evolutionary path from one species to another. They usually do a pretty decent job predicting where along our evolutionary pathway each species may have appeared! A genuine benefit to using the skulls for these kinds of activities is that they give students physical, tangible evidence of our ancestry. I have found that even students who adamantly disagree with the idea of evolution cannot help but drop their guard when presented with solid evidence such as these skull replicas.

The biology department at my school decided on the following sequence for our tenth grade biology courses:

This series of hominin skull replicas is on display in the STEM Center at Russellville High School in Russellville, Arkansas.

ecological interactions, biochemistry and carbon compounds, cellular biology and energetics, molecular genetics and heredity, and evolution. However, we all make sure that evolution as a concept is the underlying theme through which all the other content is presented and explored. For instance, as students begin learning about ecological interactions such as predator-prey relationships, they are introduced to the concept of an evolutionary arms race. I may ask a bell-ringer question about cheetahs and gazelles, and ask students to consider how cheetahs could become so fast or gazelles so agile. They usually speculate that the fastest cheetahs or the most agile gazelles were the ones who could

survive best, and thus they independently come to the conclusion of natural selection without any real guidance on my part. I really enjoy approaching concepts in ways that allow students to discover the information themselves, and evolution really is such an obvious concept that it is perfect for students to explore and hypothesize about independently.

Favorite Investigative Unit or Activity

As Darwin concluded after his voyage around the world in the mid-1800s, nature is the ultimate decider of the winners and losers in the race for existence. That race results in gradual descent with modification, and this process is probably one of the hardest for younger students to wrap their minds around. I would wager it is difficult even for fully cognizant adults to understand, including those of us who may attempt to teach it: that one species can gradually become another, and that there was never a "first" chicken, or indeed of any species. The University of Utah's Genetic Science Learning Center has developed an amazing free curriculum for teachers to use to teach evolution. The curriculum begins with an introduction to genetics via shared biochemistry, then moves into common ancestry, heredity, natural selection, and speciation. The curriculum is chock-full of engaging activities, but if I were to try to narrow it down to one thing that I felt would help students understand evolution the best, it would be their "Fish or Mammals?" case study.

This case study can be found on the TIES website. In it, students use evidence to solve the mystery of whale ancestry. They consider evidence from anatomy, fossils, and genetics to determine how modern whales came to be through a gradual, step-by-step process from terrestrial four-limbed ancestors. The case study exposes them to a variety of lines of evidence that all point to the same conclusion. These lines of evidence are often taught disparately in science classrooms, but what I love about this case study is that it brings them all together to tell a singular story: organisms share common ancestry.

Dealing with Conflicts

I teach biology in Arkansas, a state that has dealt with more than its fair share of challenges to teaching evolution in public schools. According to the National Center for Science Education, the 1968 *Epperson v. Arkansas* decision established that the state could not prohibit teaching evolution, saying that ". . . the US Constitution does not permit a state to require that teaching and learning must be tailored to the principles or prohibitions of any particular religious sect or doctrine." A short time later, in 1982, the decision in *McLean v. Arkansas Board of Education* found that the state's rule allowing equal time for "creation-science" and "evolution-science" was unconstitutional since "creation-science" isn't actually a science and the statute didn't have a secular purpose.

These kinds of religiously motivated conflicts are certainly not unique to my state, but that sort of conflict is the most pernicious one I face nearly every year. The current school where I teach is actually surprisingly diverse for the area, with a fairly high minority population both in race and in religious identity or lack thereof. Despite this,

Chance Duncan discussing reptiles at Arkansas Governor's School in July 2021, holding an eastern indigo snake (*Drymarchon couperi*).

I all too frequently hear about local churches who bring in "science experts" who promote creationist propaganda and encourage youth congregants to go back to school and challenge their teachers about scientific concepts such as evolution and deep time. While many religious individuals find no problem accepting what we know about evolution and the age of the universe, it is impossible to teach young-Earth creationists real science without offending them.

I'll close my chapter with a poem (of sorts) I created to highlight the amazement I have felt studying how living things came to be, all the way from that first cell to us.

My name is Joshua Chance Duncan, but I go by Chance.
I am human.
Homo sapiens
A member of the last remaining hominin
Though, doubtlessly, not the last iteration of this great
    lineage.
A history, stretching backward,
Unimaginable, unfathomable, yet experienced.
Endured, successfully, by every ancestor.
Not just my father Bill.
Not just my mother Teresa.
Not just my grandparents JoAnn, Ira, Carolyn, and
    Terry.
Countless interactions, any one of which could have
    ended the lineage.
Every. Single. Ancestor. Survived.
An unbroken thread, stretching backward.
At the end I stand, holding the legacy of the lineage.
I would have to walk some 3.5-4.5 billion years back,
Just to reach my oldest ancestor.
An unbroken thread, stretching forward.
From the first self-replicating organic material.
Molecules, unknowingly competing for survival even
    then.
Wrapping themselves in lipids and being the first cells.
Continue competition.
Winners, and losers. But not one of my ancestors, the latter.
An unbroken thread, touching the first cell, then its
    progeny.
Continuing forward, touching every individual,
A long line of descent with modification.
Single cells work together, no longer competing to the death,

Cooperating, even in engulfing each other.
Endosymbiosis: chloroplasts, or in my ancestor's case,
    mitochondria.
Cooperating, into multicellularity.
Amoeba-like becomes animal-like.
Aquatic environment streamlines, and makes fishes.
In my embryonic state, I briefly looked like you.
Competition, driving individuals to exploit new niches,
And make amphibians. I thank you for my tetrapod
    features.
Harsh terrestrial environments, desiccation is a worry.
So we became reptiles, and thanks for no longer
    requiring water for reproduction.
Changing climates challenge ectotherms, so endothermy
    it is!
Some cousins shift scales into feathers and take flight.
My ancestors shift scales into hair, and evolve mammary
    glands.
We still laid eggs, hadn't figured out internal incubation
    yet.
But then, we did.
To the trees for safety! Grasping hands helped us climb.
But then, drying lands killed the trees, and grasslands
    abound.
Bipedalism, awkward and clumsy at first, freed our
    grasping hands!
Height allowed us to see: food and foes.
Hands made tools, tools caught meat, meat grew brain.
And brain size doubled, tripled.
Consciousness . . . not sure when it happened. Irony.
But at some point, my ancestors became self-aware.
It was like waking from sleep, looking back, wondering
    about origins.
An unbroken thread, held in the hands of every
    ancestor.
So many opportunities for it to break, a connection to
    not be made,
A foe triumphant in battle, or in seeking dinner.
Even the smallest break and I would not be here.
Yet here, I am.
It is 2021, or some 3.5-4.5 billion years after my earliest
    ancestor.
And here I stand, holding the end of an unbroken thread,
Stretching back, unimaginably distant, to that first
    successful self-replicating molecule.

We Are All Cousins

7

Reginald Finley, Sr., PhD
Orlando, FL

It was Day One of biology class at a local high school in Florida. I introduced myself as Mr. Finley and exclaimed that I absolutely loved science and boldly proclaimed that they will too by the end of the school year. I had my students fill out a short questionnaire about themselves asking them their nicknames, favorite foods, where are they from, favorite sports and teams, hobbies, what career do they envision themselves in and lastly, what did they think they would learn in biology this year. For the last question, their responses included animals, anatomy, medicine, and zoology. I responded how awesome that all is and that we *will* indeed touch on those subjects. But I also told them my biggest goal was for them to see the world from a fresh and different perspective and to discover things about themselves and their reality that they never considered. One student asked, "Like what, Mr. Finley?" I began to discuss the demodex mites that were probably crawling around in the roots of their eyelashes and that you can sometimes feel them moving around their eyes and face at night. I showed them some electron micrographs to give them nightmares. Yes, this was just Day One. The purpose, of course, was to prime the students for excitement and adventure. It was successful. I piqued their curiosity, and they wished to learn more. The hooks were in. The question then becomes, how to sustain this level of curiosity and anticipation.

The Organism of the Day

I created a daily organism segment for my students in which I would show them PowerPoints with crazy facts and video footage of rare, cute, unknown, weird, dangerous, and sometimes extinct creatures. It was so successful that I was able to use it as a moderating tool for behavior. The organism segments were generally either before or after the lesson depending on behavior and not at all if they were not focused on their work for that day. It was difficult not to snicker when you'd hear the jocks yell to each other to be quiet and get to work so that, "Mr. Finley can do the Organism of the Day!" Most days I tried to make sure it was relevant to the day's lesson, but in many cases, it was just good fun.

The Genetics Unit

To prepare for Evolution Week, and yes, that is what I called it, I knew that it was prudent to first discuss genetics. This is common in most biological classrooms, but I knew that I would want to make sure that the lesson was on the right trajectory to show the connection between genetics and evolution. Like many science teachers, after instructing students in dihybrid crosses and going over polygenic traits, I moved into sex-linked and autosomal traits and the students had a great time constructing their pedigree charts. Students found themselves busily working out what traits they received from what parents. Students with only one known biological parent were able to conduct possible phenotypic traits that their missing parent may have had. My adopted students were also loving it as I had them either do their adoptive parents' traits or they created a "ghost" image of what their possible biological parents were like based on their current phenotypic traits and sex-linked and autosomal traits.

We Are All Cousins

The pedigree charts are how I lead my students in the understanding of phylogeny. As we are discussing traits from our ancestors, I inquire, "OK, class! What does it mean then if I have unique genetic markers that only two students in the entire school have?" The students are silent at first. I follow up with, "Well, I had to get them from someone? What does this mean, class?" A curious student slowly raises his hand and questions, "From our parents?" Indeed, I responded. "But if I get these unique genetic markers from my parents and they got theirs from their parents, what does this mean?" Finally, more hands went up. They got it! I showed them a snippet of my family tree, but I arranged it in an unfamiliar way. Of course, this is a kind of phylogenetic tree they would soon come to learn about and understand later. I highlight an ancestor and I explain that this particular ancestor may have a unique trait that is passed on to someone down the line and that if that trait is useful that it just may help aid in the survival of the next generation. "It's like passing down super powers!"

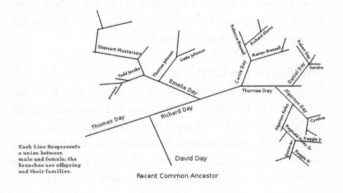

Each Line Respresents a union between male and female, the branches are offspring and their families.

Recent Common Ancestor

Image Source: Reginald Finley, Sr., Familyoriginstree.com (2015)

I went further to explain that we all share common markers between each other. I inquired again, "What does this imply, class?" Someone yelled out, "We are all related?!" I responded, "Yep! We are all cousins! Some of us are more closely related to each other than others, but genetically, we all share a common ancestor." To which he replied, "Man, that's crazy." He then turned to his desk mate and said, "What's up, cuz!" I challenged my students to remember that day to day, that we are all cousins within a grand tree of life. From then on and until the end of the school year, students were calling each other cousins. Of course, I'd love to point out to the couples in the class that they are cousins too. Everyone except the couples found that hilarious.

To continue to pique their interest, I would spend some time sharing with them amazingly strange and weird mutations that some humans actually have and that these mutations could possibly benefit the humans species if the environment dictated it. We discussed harder bones, stronger and larger muscles, and UV-resistant skin. The students appeared to now understand how beneficial and deleterious genes are passed on and what that meant for the survival of family's genes. I figured that we were now ready to move into the evolution portion but not quite yet!

My Journey To This Point

Before I get into the guts of the evolution instruction and what I think worked well for me, I'd like to provide some background about who I am and why I got involved in teaching the biological sciences. I was fortunate in that I was raised in two different secular homes. You see, my mother and father divorced only three months after I was born. I lived between both homes, spending a few years at each. My brothers left to join the military, and I lived mostly with my mother. My mother never took me to church but offered to take me if I wished to go. Instead, I opted to go with friends. My friends came from many different religious and cultural backgrounds so out of curiosity, I attended many of their services. I was able to witness dozens of rituals in temples, mosques, and churches. I remember thinking that some people have the most oddest of beliefs. My childhood friends, Anthony and Dominick, were from a Catholic family whose father believed he could dissipate clouds with his mind. I met a woman who claimed she could see angels outside of her window. I was terrified. My best friend's grandmother believed that her god told her when it was going to snow, but oddly, not when it rained. Though I was very interested in these things, I was highly skeptical

VIDEO RECOMMENDATIONS

The Natural Selection video by Stated Clearly is one of the best short videos on this topic that my students say helps them understand the topic well. Stated Clearly has other videos that are effective as well such as "What is Evolution?" and "What is the Evidence for Evolution". The Organism of the Day sessions really helped students better understand the examples provided in the videos about how gradual change over time leads to speciation.

of it all. The inexplicable nature of the supernatural I found quite disturbing.

I was curious about everything growing up, and I was fortunate that I had a mother and father that allowed me to explore my ideas. I remember asking my mother questions about this or that belief or a particular scientific idea, and she would always remind me to look it up in our *World Book Encyclopedia* collection. My mother was a huge fan of literature and had hundreds of books from the *Reader's Digest Condensed Books* collection. I never took an interest in fictional works of literature and leaned heavily towards non-fiction. I remember when my collie puppy chewed up most of the *World Book Encyclopedias.* My mother was livid. A few years later, she purchased a brand-new set of encyclopedias from *Encyclopedia Britannica*. I remember thinking how beautiful they were. I truly felt as if the knowledge of the universe was suddenly revealed to me. Much like with the internet today, I was dumbfounded at how there could be so much ignorance in the world while encyclopedias existed.

In high school, I leaned toward the sciences and music. I was much better at singing than I was in understanding science. I dropped out of chemistry and a few years later, I took a microbiology course from the same teacher who taught my brother ten years earlier. I failed it. It was then that I realized that science is hard and was probably not my calling. I loved to talk about science, but the details were often cloudy and confusing. I left singing and science behind and joined the military. After my service in the US Army, I went through phases of pseudoscience and religious beliefs but I was never fully indoctrinated. I remember always asking myself, "How do they know this is true?" In college, I took a few religious and philosophy courses and it was then that my mind was opened. Not so

much that my beliefs were wrong, which they were, but that all things should be questioned critically and unapologetically. I began to see power in questioning. Once on the path to free inquiry, I slowly left religion and pseudoscience behind. This was a natural consequence, of course, because questioning the faith was definitely not something that was encouraged by the belief's doctrines. The way my mother raised me, my college experiences, and my rational mind could no longer accept things without adequate evidence. As science is all about inquiry, my love for science was reawakened.

With my newfound critical faculties, I began to seek out others that had a passion for science and discovering reality. I discovered the Secular Web and from there I learned about the many popularizers of science and philosophy. I sought out streaming media programs critical of pseudoscience and unfalsifiable beliefs. I created a few web pages criticizing pseudoscience, and I even worked as a psychic with the Psychic Network to gain an inside understanding of the kind of people who call into these hotlines. In 1998, I eventually discovered the now defunct Atheism, Freedom, and Liberation show (AFL); in which Jake, its host, politely showed me how to create my own online media program. Shortly after, the *Infidel Guy Show* was born, lasting twelve years. The program focused on challenging unsupported beliefs, whether they were based on pseudoscience, religion, or any other idea not supported through science. I interviewed hundreds of scientists, philosophers, civic activists, ethicists, and theologians. My favorite programs were the ones in which I interviewed biologists. I interviewed Richard Dawkins, Neil Shubin, Massimo Pigliucci, Mark Vuletic, Jerry Coyne, PZ Myers, and so many other popular

BOOK RECOMMENDATIONS

The book I would recommend is Jerry A. Coyne's book, *Why Evolution is True* (2010). In an excellent YouTube video posted by the Richard Dawkins Foundation for Reason & Science, Coyne touches on many of the points in his book. He covers scientific methodology, what is evolution, geologic evidence, fossils, anatomical structures, genetics, and human evolution.

icons of evolution education. They provided me with the knowledge, tools, and confidence to be more vocal about science and evolution. It wasn't until I read Daniel Fairbanks's *Relics of Eden* (2009), that everything really clicked for me on a foundational level. I felt that I had uncovered new tools in my pedagogical repertoire to finally remove the veil of doubt in the evolution deniers.

Fairbanks wrote of the evidence for evolution via pseudogene homology, the vitamin C broken gene, endogenous retroviruses, teeth appearing in animals that don't have them in adult life, missing and fused chromosomes, and the phylogenetic tree. I knew that this information was going to become part of my evidence for evolution. After my mother passed in 2011, I stopped criticizing religion and I began to focus my efforts on science education and critical thinking. I found those aspects far more valuable to the advancement of society than attacking emotionally laden, unfalsifiable personal beliefs.

Spreading the Word

I began to share my evolution toolset with the public. I appeared on radio programs, podcasts, participated in debates, and submitted commentary to various newspapers. I naively assumed that others, even those with different attitudes about human evolution, would begin to see things differently. Instead of listening to the evidence, I was denigrated and accused of being against God or being an atheist. The resistance to evolution was quite a shock to me, so I began formulating ways to make evolution more comprehensible and palatable.

Becoming an Educator

I decided that I wished to become a formal educator in the sciences, so I went back to school to earn my first degree in science. I earned a bachelor of science degree in human development which gave me a foundation to better understand biological human development and its intersection with sociology and psychology. However, I was concerned that I had no formal training in science education and communication so I sought to earn my masters degree in the public understanding of science at SUNY Buffalo. I had the opportunity to study with Professor John Shook, whom I had interviewed many years prior. and the legendary Benjamin Radford, a scientific investigator and skeptic. During my studies there, I started my first science education business, The Fun Scientists LLC, in which my crew and I put on various physical science programs for children. I eventually sold my business assets and moved to Orlando, Florida because I was offered a position as the education director for a new science museum that was opening up. Later I was given the opportunity to teach formally full-time in a school setting. My first position was to teach biology in an Orange County, Florida, public school. The first paragraph of this chapter begins there. To further my biological knowledge, I enrolled in a graduate program in the biological sciences at Clemson University and earned my masters degree in 2016. During this time, I began working for a small private school as its advanced sciences educator and was promoted to the *science department chair*.

My Research Project

While earning my master's degree in biology, I had to take a scientific research course. In this course, we were to practice analyzing and conducting research. For one of my projects, I set out to see if knowledge and attitudes about evolution can change if students are provided proper evolution instruction in an environment that is fun, open, and socratic.

Many studies have been published on the public's understanding of science in general, but few have measured students' knowledge about evolution or explored their beliefs about evolution after they were instructed adequately in high school. I decided to survey my class, which I usually do anyway after every lesson but I wanted something a little more informative with regard to this particularly controversial topic. I was hoping to discover how I was doing as an educator on this particular topic and to learn what students actually thought about evolution as opposed to what they actually learned. I needed some effective and accurate material with which to educate my students. This is when I considered help from the Teacher Institute for Evolutionary Science (TIES). Bertha Vázquez soon provided the teacher's version of the TIES Evolution unit. I customized it for my students, and I finally felt prepared.

Evolution Week

I first asked my institution if it was OK to survey the students for my university project, and I was informed that as long as it was not "formerly published" and anonymous that it was OK. I went on ahead and received parental permission anyway. Luckily, there were no issues and all students were given the okay to participate. I provided the students a twenty question survey reflecting their acceptance of evolution as well as assessing their current knowledge about evolution. A Likert scale survey was used in which students had to answer whether they: strongly agree, agree, neither, disagree, strongly disagree. The middle option of neither agree or disagree was for the student to simply not select an option. The survey was given using Kahoot.it,

an online polling and quizzing tool. A snippet of one of the questions can be seen below followed by the complete list of statements to which the students responded.

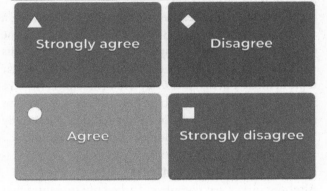

	Items
Q1	Biological evolution states that humans come from monkeys.
Q2	Evolution states that organisms are constantly improving.
Q3	Survival of the fittest means the stronger and bigger organisms win.
Q4	Organisms evolve purposefully to satisfy biological needs
Q5	Evolution attempts to explain how everything came to be.
Q6	Scientists have never seen a species evolve.
Q7	Evolution is just a theory.
Q8	Evolution doesn't make any sense to me.
Q9	Evolution is just completely random.
Q10	In general, I think I understand the theory of evolution.
Q11	I accept the scientific theory of biological evolution.
Q12	I accept the scientific theory of human evolution.
Q13	My views on the acceptance of evolution is due to scientific consensus.
Q14	My acceptance/non-acceptance of evolution is influenced by unscientific or religious beliefs.
Q15	My religious/personal beliefs do not stop me from accepting evolutionary theory.
Q16	I accept evolution for other animals, not humans.
Q17	I do NOT accept the theory of evolution in any fashion.
Q18	I do NOT accept it because I do not understand it.
Q19	I do NOT accept any form of evolution due to personal or religious reasons.
Q20	I do not accept the scientific theory of evolution due to other reasons.

https://tieseducation.org/book

I was concerned that the students would answer the questions in a way to please me, but I encouraged them to be honest. The topic of religion had come up before in my class, and I just informed them that my class was a science class and that science would therefore be the focus. That said, however, I also informed them that I really don't care what anyone believes as long they aren't harming anyone. The students knew that I was nonreligious, and they seemed OK with that. I also felt that the students would be mostly honest in their responses because it was late enough in the year, and there was a built-up rapport. The students knew that I was always open to their ideas and beliefs. They really liked that it was anonymous as well. They were informed that the survey was optional. It was just for me to evaluate how they feel about the course topic and to see if I am teaching evolution properly.

I convinced a colleague of mine who was teaching a section on education history to give his students the same survey so that I could compare the means of the results with his class. The Scopes trial became quite a popular topic of interest in his class. Only a few students in his class had already taken biology, but they had taken it last year from another teacher who openly stated that she literally "hated" plant biology and doesn't "believe" in evolution.

Evolution Misconceptions

After the survey was complete, I asked my students what they thought evolution was. The answers varied but the top five most common ideas were that organisms make themselves change for their environment, that humans come from monkeys, evolution is just a belief or theory, the strongest survive, and that evolution explains how all life got here. I decided to have a little fun with the students and addressed each of these questions via humor and Socratic methodology.

With regard to the first idea that organisms can just make themselves change to suit the environment, I discussed that this is not Pokémon. They laughed, and I assured them that evolution just doesn't work this way. I informed them that the environment could change right now, and it would be wonderful to have wings to help us survive. But we don't have genes for wings, and no human has ever been able to spontaneously produce a limb anyway. Therefore, if the environment changed

too quickly and having wings would be the only way we could survive, we'd all just die. No Poké Balls to save us. To address their idea that humans come from monkeys, I put up an image of a ring-tailed lemur, a common chimp, a lion-tailed macaque, and me on the whiteboard. I asked the class to tell me which of these animals do they think share more DNA. I started with me, and then the students all generally worked the genetic distance between them solely based on homology. The results were: Mr. Finley, chimp, macaque, and lemur. I then proceeded to inquire about the origins of the lemurs and they remembered the family tree revealing cousin relationships. They understood that the macaque and lemurs must be cousins and therefore share a great-great-great grandparent. So I asked them if the lemurs came from the macaques based on the tree I was placing on the board. They quickly realized that this couldn't be case. We continued on with the chimp and macaque and then between chimps and humans, and they now understood. That we are cousins with chimps, not descended from them.

The definition of a scientific theory was discussed during the first week of class so I was surprised to see so many students referring to evolution as "just a theory." I asked my students whether something can be a fact and a theory. They weren't sure. I asked them, "Is gravity a fact?" They quickly answered in the affirmative. I said, "Yes, of course it is. But did you know that the explanations for how gravity works are theories? This is the same with evolution. It is a fact of reality as observed via the evidence. The process of how it works is evolutionary theory."

The idea that the strongest survive was a challenge I decided to put back on the students. I asked them to provide me an example of how the strongest is not the fittest. A few students quickly pointed out disease, accidents, and gun violence as examples. They were exactly right! I think some of the students were simply stating what they believed "fittest" meant when they answered "strongest." I also challenged them to give me examples where strength can benefit a species, and they mentioned elephants and tigers. Indeed, having such a power to defend or kill may help a species survive.

Lastly, we touched on that idea that evolution is a theory that explains how all life got here on Earth. My deeply religious students appeared almost relieved when I informed them that evolution does not discuss the origins of life. Only that once life took a foothold, that organisms flourished through gradual biological changes. With this, I think the students were now ready to engage deeper into the lesson.

Including discussions, labs, and supporting videos, the evolution course took us eight days to complete. The TIES presentation included these topics: fossils, biogeography, the law of superposition, artificial selection, vestigial organs (the evolutionary legacy we carry within our own bodies), similarities (comparative anatomy), the overwhelming genetic evidence, and the five principles of natural selection (that individuals in a population show variation, variations can be inherited, organisms overproduce, variations increase reproductive success, and that populations will slowly change in response to the environment). To sweeten the pot of information, I used Fairbanks's (2009) information on pseudogene homology, the vitamin C broken gene, endogenous retroviruses, teeth appearing in animals that don't have them in adult life, and missing and fused chromosomes. I reintroduced the phylogenetic tree alongside with bacterial resistance and the classic peppered moth example of how genetic change aids in the survival of a species. The last slide I added was the predictive power of evolution in which I discussed the Morgan's sphinx hawk moth, which was predicted by Charles Darwin 130 years earlier.

As I mentioned earlier, I found it vital to open up the floor to questions and answers as we moved through the program. Some students were confused and stated that they are not allowed to consider evolution in their homes or churches. Others stated that they don't know what to believe. I referred them to videos featuring Jeffrey Selman and Kenneth Miller discussing faith and science on YouTube and videos on critical thinking. Some students were relieved that they could have their faith and accept the evidence for evolution. Some students informed me after the course that they were forever unwavering in their position. I stated that was completely fine.

At the end of the lesson, my students took the survey again. The history teacher gave his students the same survey again as well. When I ran the analysis, there was indeed a significant difference in their knowledge gains of evolution and a significant difference in their attitudes towards accepting evolution. There were lower gains in acceptance versus gains in actual knowledge however. The control class showed no significant changes in their prior positions with only a very small uptick in knowledge. That was likely because some students spoke with other students and changed their position after some reflection. Regardless, there was definitely a significant difference between the two groups and significant differences in attitudes of acceptance before and after the lesson.

The data revealed to me that at least short-term changes in knowledge and acceptance of evolution can occur if given the proper environment. To really be successful in the high school classroom, that you must have a teacher who is passionate about the material and open to allowing students to express themselves. The teacher should use Socratic methodology, employ adequate teaching tools covering misunderstandings about evolution, and anticipate resistance and challenges to evolution, in particular human evolution, which was doubted more than just evolution in general according to the survey.

The data also helped me realize where I can improve as an educator. Questions to consider were: What did I not teach correctly? How can I reduce threatening feelings to their deeply held personal beliefs? Am I teaching the right material? Is what I teach important for them to even know? How can I make it even more relatable? These questions are important when devising lessons on evolution, and you most assuredly will have even more if you are doing it correctly.

Student Reflections

I invited my students at the end of the biology course to express what were the best parts of the class. Overwhelmingly, the top three elements mentioned were:

1. Teacher passion and energy. Studies continue to show that teacher passion for the topic motivates and inspires students.
2. Relevant content. Discovering new things about their bodies and nature. The students were motivated to come to class everyday because they knew that they'd learn some thing about themselves and their world everyday. Boringness is lessened when it becomes all about them.

3. Likeability. The reality is that, by and large, they simply liked me as a teacher. Studies have revealed that disinterest and dislike in the teacher lessens opportunities for learning.

Evolution education doesn't have to be a daunting and sterile process. Anxiety that some teachers may feel teaching this subject is simply because they may not have been prepared. Luckily, for the science teacher, the evidence for evolution is ubiquitous and many educational resources are available to educate teachers and well as their students. The experience should be fun and enlightening for the student. If the students are not having fun and enjoying the process of learning, something is broken. The reality is that, usually, it is us, the teachers. We should make the lessons relevant, we need to engage with them Socratically, we should laugh, have fun, and remind them every single day that we are all cousins within this grand tree of life.

Introducing Evolution

Earlier in this chapter, I mentioned that I created the "Organism of the Day" segment to inspire and keep the students interested in biology. Revealing their cousin relationships to each other was the hook that provided an appropriate segue to the evolution lesson. The TIES Evolution Instructional unit allowed me to move through the content in an organized and effective fashion. In the unit, an image shows the cousin relationships between humans and cows, birds, cats, dogs, and other creatures. Students are shocked to see that we are more genetically similar to rats than cats! To help cement these concepts, I reveal the phylogeny of humans. I am then able to blow up the tree of life and show how we are related to all life on Earth.

Bellwork each day during our Evolution Week consisted of questions that students should reflect upon and research such as:

- How do bacteria adapt to antibiotics?
- How did dog breeds come about?
- What is a viral strain?
- Why do we look like our ancestors?
- Where does red hair come from?

Students can organize and pick a topic to quickly discuss and research. I offer bonus points for 100 percent engagement. I then ask each group to explain what they discovered.

Favorite Investigative Units or Activities

The online activity "The TIES Time Machine" is an excellent resource to allow students to use digital resources in a fun, interesting, and at times frustrating way. The activity explains what natural selection is, includes an interactive game, and a quiz to test understanding. Critical questions can be asked during the session to discuss what aspects may improve or hinder your organism's chance of survival. The game reveals that diversity is key and that a lack of diversity may mean death for a species if the environment changes too rapidly.

OneZoom features an interactive map of the evolutionary links between all living things. I can task students to find cousin relationships between branches and to make predictions about what features should be common between clades before zooming into the tree. This activity provides students an understanding of the predictive power of evolution because students can anticipate modern phenotypes between clades. We can also review the fossil record and predict what phenotypic features could migrate from ancient clades up until today.

Dealing with Conflicts

I use the student's challenges to evolution as a way to remind students of what science is and what scientific methodology is all about. They are reminded that science is a self-correcting discipline and that more science will overcome bad science, not less. Because of this, unscientific ideas are not a challenge to disproving scientific methods and discoveries. I find it important to also inform students that apparent science conflicts don't necessarily mean that one's faith is wrong, "It's just not science."

8

Cultural Border Crossing

Katie Green, PhD in Science Education
Athens, GA

As a child in the South, I grew up going to church every Sunday. I have always been a question-asker, and remember asking in Sunday School, "But where were the dinosaurs in the Garden of Eden?" I don't remember what the answer was, but I must not have accepted it, because I kept wondering. When I went away to college, I stopped going to church, not because I necessarily stopped believing in God, but because I spent Sunday mornings sleeping late and going to brunch at the dining hall with my friends. In an ethnohistory class my junior year, my professor was talking about Greek myths. I remember asking her, "So did the Greek people believe that these stories actually happened? That Zeus popped Athena out of his head?" She said, "Yes, although we call them myths, they believed these stories to be true." I was astounded, and remember thinking in my head, "That's crazy! But wait a minute . . . God made Eve from Adam's rib?!!!?!?" At that moment, sitting in a corner classroom on the quad, I realized that the stories I had read in the Bible were just that, stories. I began thinking about religion through an anthropological lens, and thinking about what purposes it served in cultures.

Life went on, and I finished bachelor's and master's degrees in anthropology. I worked as an archaeologist and was offered an adjunct position at a community college, teaching anthropology. The dean wanted me to teach a class called Human Origins. I remember a surprising phone call with him. He requested that I teach it as a Wednesday night class because "then the church people won't come." I agreed, although it had not really occurred to me that anyone might have a problem with hominid evolution. To me, science and religion both made sense. They both answered questions about life, but one was based on evidence while the other was based on faith. Most of the Human Origins students were teachers taking the class for recertification credits. Only one student had an issue with evolution, and I remember him calling me to talk about it. I promised him that he could take my class and keep his religious beliefs. That was the end of that conversation. No one else ever questioned hominid evolution, at least not out loud.

After a few years, I became certified to teach middle school science and began my first year teaching in Portland, Oregon. The first year, I taught evolution as part of a life science unit. Science teachers were encouraged to refer to evolution as "change over time." I would start the unit with a disclaimer that "evolution only means change over time," which seemed to satisfy the students and parents. Since Portland is not a bastion of conservative religious thought, I didn't think too much about how learning about evolution might affect any of my religious students. I was very surprised a few years ago, when a student I taught in 2004–2005 mentioned on Facebook her memory of being uncomfortable with learning about evolution in my class, and how I allowed

BOOK RECOMMENDATIONS

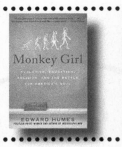

For a bit of history about teaching evolution in public schools, I recommend *Monkey Girl: Evolution, Education, Religion, and the Battle for America's Soul* (2008) by Edward Humes.

her to sit in the hall during the lesson. I was floored by this comment. I allowed her to sit in the hall? Did I give her another assignment? Did she feel cast aside because of her religious beliefs? Her memory of my solution to the tension she felt seemed positive, but I felt sure I could have behaved in a more accommodating manner.

Over the years, I continued to teach anthropology at the community college level as well. Hominid evolution is a major component of any anthropology class that includes a biological anthropology section. I noticed how many students misunderstood evolution. I would see discussion board posts which said things like, "Evolution sounds like an interesting theory, but unfortunately, I don't believe in it," or "Scientists today are still debating whether evolution is true anyway." There seemed to be an issue with students' acceptance of evolution as a valid scientific theory. Some students articulated an ability to mesh their religious beliefs with scientific views on evolution, while others thought one had to choose between science and religion.

When I started my PhD in science education at North Carolina State University, two big factors nudged me in the direction of evolution education research. First, I took an elective called evolutionary ecology. I connected right away with the professor, Brian Langerhans, who gave extra credit when you could name the rock star who sang the lyrics associated with the topic of the day. (For example, in Brian's mind, the lyrics from Bruce Springsteen's "Thunder Road" are related to constraining components that shape evolutionary trajectories.) Most days I listened and thought, "Did I teach this right? Evolution was so much more complicated than what I had taught in middle school science, or the quick review I offered before learning about hominid evolution. I focused on natural selection, but what about

sexual selection? And epigenetics? All of this thinking led me to write my National Science Foundation Graduate Research Fellowship Proposal (NSF GRFP) about how evolution was being taught in community colleges. I received honorable mention for the proposal, which put me on the path to researching evolution education for my dissertation.

A second, more subtle nudge, took me a while to notice. During my first year as a doctoral student, I had a myriad of responsibilities, including my classes, researching arts integration into science classes, writing my GRFP, and supervising student teachers. I also had several research interests other than evolution education such as teacher retention and field trip pedagogy. My advisor was encouraging me to make some decisions about my research path. I realized that every time I entered my office with a list of things to do, I started with the evolution education task first. Evolution teaching and learning were my passions, even if it took me a while to admit it.

As I moved forward with my dissertation research, I became more interested in the relationship between science and religion. My "polite Southern girl" side experienced anxiety when I even considered talking to people about controversial subjects. I kept saying, "I want to study evolution, but I don't want to do any research on the religious arguments against it." As an agnostic, I didn't want any of my religious friends or family to think that I was critical of their religious beliefs when I talked about my research. Evolution just seemed so charged. Finally, in my reading, I found two articles that alleviated my anxiety about evolution education research.

In his work with First Nations people in Canada, Glen Aikenhead theorized that some students might undergo *cultural border crossing* when moving between

VIDEO RECOMMENDATIONS

A favorite video is *Natural Selection with the Amoeba Sisters.* It provides clear explanations of evolution and natural selection along with other often-confusing terms like "fitness."

their home culture and the culture of the science classroom (Aikenhead 1996). While his work discussed science in general, not evolution, I realized that American students might also cross cultural borders when they move from a religiously based home into a biology class where they studied evolution. At the same time, I read another article which discussed Jegede's theory about collateral learning, the result of cognitive conflicts that emerge from differences between students' cultural beliefs and science classes (Aikenhead and Jegede 1999). These theoretical frameworks allow for acceptance of both scientific and religious explanations about life. Finally, this felt right to me. I could talk to people about how their religious beliefs conflicted with evolution if I didn't have to advocate for discarding their beliefs.

Thankfully, I found a community college instructor, Mr. Gloucester, who was willing to try a cultural border crossing intervention in his Introductory Biology class's evolution unit. Using the cultural border crossing and collateral learning frameworks, I designed an intervention to see how it could affect students' understanding and acceptance of evolution. I had originally planned to create this intervention for a university biology class. Since Gallup poll results indicate 38 percent of people believe humans in their current form were created by God (Swift 2017), I decided to conduct my research in the community college context hoping it was more representative of the general public.

On Day One of the intervention, we focused on multiple epistemologies, or sources of knowledge. To begin, we gave groups of students a list of questions such as "What happens to us after we die?", "When did humans appear on Earth?", and "What is a cell made of?" The students had to create categories of questions based on the sources of knowledge they could use to answer the questions. Some students created source of knowledge categories such as "scientist," "God," and "the Law," while others divided the questions into only "Philosophical" and "Science." Only one group of students placed the questions between two source of knowledge categories. While I had hoped that more students would realize that some questions could be answered by two sources of knowledge, this activity was still promising as an introduction to the idea that people answer questions with different sources of knowledge. Mr. Gloucester concluded by talking about the idea of multiple epistemologies. He asked the students if they felt like they lived in two different worlds. While some students answered yes, they were not willing to elaborate.

Day two focused on creation stories. Mr. Gloucester introduced the topic by reading the creation story of the Haida, a tribe on the Northwest Coast of the US. The story is of Raven, who discovered the first men in a clamshell along the edge of the surf. After sharing this story, the students were grouped and asked to find another culture's creation story on the internet and share it with the class. Students found lots of interesting stories, varying from Hindu origin stories to ones from Australian aborigines. One of the aboriginal stories involved a rainbow serpent coming up through the ground. A male student in the back yelled out, "That's crazy!" It reminded me of my own experience thinking about Greek myths in the ethnohistory class so long ago, and I wondered if this student was having a similar experience. After sharing the stories, Mr. Gloucester asked, "Can a Haida believe their creation myth *and* a scientific account of how humans came about? Or do they need to give up their religious or cultural beliefs to accept scientific evidence as valid?" We were hoping students would agree that people could hold collateral knowledge, as Jegede posited. At least one of the students agreed that the Haida could believe in both accounts of how humans arose on Earth.

On Day Three, each student pair was assigned a religion and was asked to investigate the doctrine about evolution associated with the religion. I created an answer key from the Pew Research Center, choosing the religions on which they had information. The religions included Judaism, Islam, Mormonism, and Pentacostal Holiness. As Mr. Gloucester wrote the religions' names on the board so students could fill in the chart, he remarked, "Writing the names of religions on the board is the weirdest thing I've ever done in a science class." When students filled in the chart, we could see that with the exception of Southern Baptists, none of the religions on the list specifically opposed a belief in evolution. While talking about the Catholic doctrine, the students discussed theistic evolution, a great example of secured collateral learning, when a learner is able to allow for interaction between conflicting ideas (Aikenhead and Jegede 1999). Elise, a Muslim student originally from Morocco, explained the various Islamic sects and detailed their beliefs in evolution.

While the discussion of religion and evolution was very interesting, the most fascinating moment came at the end of the lesson. As students were discussing the chart of religious doctrine, Mr. Gloucester cupped his hands around his mouth like a megaphone and loudly said, "And you know who believes that humans came from apes? *No one!*" There was an audible gasp from the back of the classroom, and one girl shouted, "*What?* But what about that drawing where the monkey turns into a human?!!?" Mr. Gloucester explained that this drawing was not intended as a literal explanation for human evolution, and that no scientists believed that humans evolved in a linear fashion from a monkey like the drawing portrays. The students were flabbergasted, and the six students I interviewed for my dissertation mentioned in the post-semester interviews that this was one of the most important facts they learned in BIO 110. Two of the very religious students I interviewed mentioned that they felt "relieved" to know that no scientists expected them to believe that humans evolved according to the famous drawing.

On the fourth day of class, Mr. Gloucester collected the School Board Scenario (Binns and Bloom 2017),

an assignment that had been distributed at the beginning of the unit, with students given a week to complete it. This assignment asked students to define science, then read about creationism, evolution, and intelligent design. Students then assumed the role of a school board member and explained which of the three ideas would be taught in the public school science classroom and why. The discussion about this assignment was quite lively. Mr. Gloucester started by asking if anyone wanted to share why creationism should be taught in the science classroom. One of the most religious Christian students, McKenzie, said, "I said it should be taught in the science classroom because they call it 'the theory of evolution' and technically creationism and intelligent design are also theories, so why shouldn't they be taught too?" Another of the most religious Christian students, Sophie, said, "I said both should be taught, too, because why not teach all three and let the students decide?" Elise, one of the Muslim students, explained, "I said no to both because according to the definitions, they don't align with science. I think it's a personal choice as to what people choose to believe. But it's a science curriculum, so it's not about what you choose to believe." Another Muslim student, Reema, said, "I said no to both because nothing related to religion should ever be taught in public schools." Neko, one of the atheist students, said, "Religion in schools has always been evangelistic. If we teach one religion's creationism story, you would have to share all the other religions' ideas." Mr. Gloucester responded that if he had to do that, his evolution unit "would take five years." Finally, Tom, a student whose religious beliefs were unknown to me, said that he would choose to include intelligent design only. He explained, "I think creationism is just stupid stories that people make up like moles and yarn and people squeezing stuff out of water and I've heard better campfire stories than all the creationism stories and the idea that the world is complex [intelligent design] makes more sense."

After students shared their opinions, Mr. Gloucester moved on to question the students a bit about the nature of science (NOS). While I didn't use this instrument in my dissertation research, research has been done on students' views on NOS using an instrument created by Lederman and colleagues. In this instrument, the views of nature of science (VNOS) questionnaire, includes

Katie Green crossing an ancient border at Pompeii.

aspects of NOS that are relevant to students: science is tentative, based on empirical evidence, theory-laden, partially based on human assumptions, and fixed within social and cultural contexts (Lederman et al. 2002). Mr. Gloucester began the discussion by asking students what kind of evidence people would need to accept science as the truth. One student responded that evidence was needed. To gauge how students thought about the empirical nature of science, Mr. Gloucester asked if scientists could prove things with only one experiment. Cooper, an agnostic student, answered that multiple experiments

were needed to prove scientific theories. Mr. Gloucester agreed, and said that scientists often base their theories on thousands of experiments, and rarely only on one. He then asked the students about the way scientists use the word "theory" as opposed to the way we use it "on the street." Students replied that the two uses were different, but did not elaborate.

At the end of the discussion on the School Board Scenario assignment, Mr. Gloucester asked why we teach evolution in science class when there is obviously not universal agreement on its validity. Elise answered that evolution is "the best theory" to explain what happened in the history of life on Earth. Mr. Gloucester added that evolution is the theory most supported by scientific evidence. Neko asked, "Don't scientists in biology and other fields depend on evolution?" Mr. Gloucester answered affirmatively and elaborated by saying, "A lot of other parts of science depend on the theory, and it holds all the parts of science together." While this was the end of the class discussion, my dissertation data analysis examined the School Board Scenario assignment further and found that students' explanations of why creationism or intelligent design should be included in science curriculum quite often contradicted the definitions of science students wrote at the beginning of the assignment.

In addition to attending class during the evolution unit, I administered instruments that measured students' understanding and acceptance of evolution before and after the unit and interviewed a variety of students before and after as well. While I have not completed the data analysis, I can see that this intervention did not completely relieve the tension between students' religious beliefs and evolution, though it did have a small effect on their understanding and acceptance of evolution. After the unit, all the students reported that they enjoyed the intervention, and I hope that they will continue to think about the possibility of holding conflicting views about multiple sources of knowledge when considering important scientific topics like evolution. All but one of the students agreed in the final interview sessions that people could hold competing ideas and somehow combine them together. Neko, the atheist student, felt that at some point, everyone would have to choose between science and religion.

My journey to studying evolution as a science education researcher has been heavily influenced by my experience as a classroom teacher of both middle school and community college students. My dissertation

defense will certainly not be the end of this journey. Completing my research only increases my certainty that science and religion aren't hopelessly antagonistic. I plan to further this research interest by investigating how students' NOS beliefs influence their thinking about evidence and faith regarding evolution and other socioscientific issues. While I don't engage in self-delusion that I can convince all students that evolution is a valid explanation for the variety of life on Earth, I believe my research in this area can assist teachers who struggle to reach religious students.

Introducing Evolution

The research discussed in this chapter is focused on how evolution is taught in an undergraduate introductory biology course. I started the unit by acknowledging that there are multiple sources of knowledge (science, religion, the law, the internet, your parents, etc.) and that sources often answered different types of questions. I hoped that by recognizing that students might use other sources of knowledge to answer important questions in their lives, they might begin to think about which questions science is equipped to answer and which questions it cannot.

Favorite Investigative Unit or Activity

I find that students are often fascinated by hominid skull casts. Sometimes I have taught near universities that would kindly loan out their skull sets. Other times, I have been able to purchase a set through a site such as boneclones.com. For a less expensive alternative, try the TIES skull comparison lab at tieseducation.org/book.

I like to have student pairs make observations about the different species and create compare/contrast charts.

9 The Evolution of an Evolution Advocate—A Lifelong Journey

John S. Mead, MA in Teaching

Dallas, TX

My interest in evolution began before I learned how to read. When I was a preschool student in the early 1970s my parents had a subscription to the Time-Life Nature Library. Subscribers received a book several times a year. Despite being too young to read, I found the illustrations of these volumes enticing. The book that captured my attention more than any others was *Early Man*. Although I could not yet read, the images of species such as Neanderthal man and *Homo erectus* and especially *Paranthropus boisei* enthralled me. They were so obviously similar to us and yet so different. That book became my "security blanket" of sorts as well as the focus of good-natured family teasing in later years. I grew up in New York City and also fell in love with the American Museum of Natural History and its fossil collections. So when Don Johanson's *Lucy* was published in the late 1970s, my middle school self eagerly devoured it. High school brought me my first formal introduction to Charles Darwin and the basics of natural selection. His adventures aboard the HMS *Beagle* and the idea that evolution was somehow a "dangerous" idea in Victorian England were immensely appealing to my rebellious teenage mind.

As I look back at my initial exposure to the early history of evolutionary thought, the coverage it received in my sophomore biology text and in class was limited to the story of Darwin and Wallace both having the idea of natural selection and deciding to have their ideas read before the Linnaean Society together on July 1, 1858. This was followed by a brief mention of Thomas Henry Huxley (who was called "Darwin's bulldog"), his support of natural selection, and his famous debate with Archbishop Samuel Wilberforce. The final mention of the controversy of evolution was a brief overview of the 1925 Scopes trial. My mind was left with the impression that after Huxley and the Scopes trial, any conflict about evolution was effectively ended. At the end of the year, we were assigned a term paper that allowed us to choose a topic in biology and explore it more deeply. You will not be surprised that I chose to dig deeper into evolutionary theory.

My research for my term paper opened up a new world that the superficial in-class coverage necessarily overlooked. First, I discovered that Darwin and Wallace were not the first people to suggest that life had evolved. In my research I encountered names such as Erasmus Darwin (Charles' grandfather) and Jean-Baptiste Lamarck, who had ideas about how life changed over time. I also discovered that Darwin's development of natural selection was not a single light bulb moment of inspiration while he visited the Galápagos Islands, but rather natural selection was the culmination of a range of influences from men such as Thomas Malthus, James Hutton, and Charles Lyell. I was also introduced to a basic understanding of the work of early twentieth-century genetic pioneers such as Thomas Hunt Morgan, Theodosius Dobzhansky, and Richard Lewontin.

In addition to expanding my knowledge of the scientists who helped form our current understanding of evolutionary mechanics, I was also exposed to the fact that evolution was not a scientific idea that was accepted by everyone. While I understood evolution and accepted it because of its detailed explanation of how life came to be as it is today, I saw that large groups of people did not share this worldview. Far from the Scopes trial being the end of debate over evolution education in the United States, I discovered that there had been a resurgence of anti-evolution sentiment that was embodied in a series of legislative attempts to either remove evolution from classrooms altogether or, more commonly, to provide equal time for creationist ideas alongside evolution teaching in science classes. My teenage mindset was baffled by these efforts to dismiss proven scientific fact in place of faith-based explanations. Despite my dismay about this situation, my life was also filled with the trials and tribulations of a teenager in a prep school. As such, my focus drifted from a focus on evolution toward sports, friendships, a new semester of classes, and thoughts of college applications. However, the seed had been planted . . .

In addition to my love of evolution, I had grown up in a medical household and also enjoyed spending time tagging along with my father during his patient visits. I was allowed to serve as an operating room gopher for several summers in high school which set the stage for me to think about entering college on a pre-med track. At this point, I had no idea that there was a field of study called evolutionary biology so pre-med it was. As I got disillusioned with the cutthroat nature of the pre-med universe, I took a shine to ecology and then in the summer after my junior year I earned an internship with the Student Conservation Association (SCA). My internship with the SCA had me assigned to the Curecanti National Recreation Area just west of Gunnison, Colorado, adjacent to the Black Canyon of the Gunnison National Park. During my internship, I took on the role of an interpretive park ranger and was tasked with telling park visitors the natural history of the area. I quickly fell in love with teaching the story of this rugged western landscape and how it had changed since it was first described in the mid-nineteenth century. That SCA internship turned the tide for me. It convinced me I wanted to go into a career that involved teaching about science. In my senior year I applied and was accepted into the newly revamped master of arts in teaching program at Duke University. I should have known I made the right decision when at my graduation, our speaker was *the* Stephen J. Gould and he eagerly told the graduating class that "The Republic needs more science teachers!" I may have been the only graduate cheering that line. I then spent the next twelve months combining graduate coursework and a full high school teaching load under the expert guidance of master teacher Gloria Perkinson.

My eighteen months of work in the master of arts in teaching program was intense, to say the least. The summer course work in education theory was straightforward, but then the combined load of graduate-level biology classes and a full load of teaching biology classes in the fall term served as a wakeup call. I learned quickly that classroom teachers had little time to explore much beyond the daily work of prepping classes, teaching, and grading (in addition to my two graduate-level classes). Learning and appreciating the demands of teaching over the course of an entire school year were fantastic benefits of my year in the MAT program, but the hectic nature of that year precluded any exposure to the existence of professional development. Given the nature of teacher training in the late 1980's, the concept of networking in any meaningful way did not exist. As such, I wound up seeing teaching as something that was primarily limited to your work in one building or on one campus.

In the spring semester of that year, I began my search for my post graduate "real" job. Given my background with independent schools, I focused my search there. Since I did not have any strong geographic ties at that time, I conducted as broad a search as possible. I connected with an independent school placement firm and encouraged them to share my availability with any school seeking a biology teacher. I was surprised and pleased when I soon had interest from more than a dozen schools. I narrowed it down to five schools based on what I knew of their reputation, location, and type (boarding vs. day school and coed vs. single sex). Of those five schools, I was invited to interview at three. All three extended offers of employment, but two of them suggested that I'd be the rookie teacher who would follow in lock step with an experienced teacher to learn the ropes from them. The third school surprised me by offering a middle school position teaching a full load of life science classes alongside the department chair who had been there for more than twenty years.

In my interview with her, she made it clear that she had grown as a teacher by having the freedom to develop her own curriculum as she taught to her strengths and worked to identify and build upon weaker areas. She told me that she'd want me to be creative and innovative. And yes, I could expect to fall on my face occasionally, but she and other department members would be there to support me and pick me up when I fell. I was excited by such a progressive and supportive growth mindset! In addition, I noticed that in the hallway next to their science lecture hall was a series of display cases with minerals, a meteorology station, and a display of three cast skulls, *Australopithecus boisei*, *Homo habilis*, and *Homo erectus*. Upon asking about this unexpected display, I was told that world-famous paleoanthropologist Richard Leakey donated them during his visit to the school a few years earlier. So now I had a school that believed in me as a creative professional *and* had some sort of connection with one of the biggest names in evolution research! Could it get any better? As it turns out it got better when this school's salary offer was almost 20 percent better than any other school and I'd also be teaching in a newly completed science building that I had toured in a hard hat during my interview.

The school I have described is where I've been teaching now for twenty-nine years—the St. Mark's School of Texas in Dallas, a 1-12 independent boys school. I have never regretted my decision to move to Dallas; St. Mark's more than lived up to its promise for me as a science teacher and evolution educator.

The first two months of my career at St. Mark's were expectedly busy as I got to know my new school and settled into the rhythm of life as a rookie teacher. As expected, my experienced colleagues welcomed me into the fold and embraced me with warmth and encouragement. In early October 1990 I was surprised to get a note (this was in pre-email days!) from the headmaster requesting a meeting with me in his office. My stomach sank as I worried about what I might have done to get "called on the carpet?" After two days of worry, the meeting time arrived and I went into his office expecting the worst. Thankfully, it turned into one of the best meetings

of my life! He informed me that he had taken notice of my good work and that I had surpassed his expectations as a rookie teacher. He then went on to explain that he had an important job for me. Thanks to parents of two of our high school students, we were to be visited by a scientist in November. He wanted me to make certain that the sixth graders were prepared for the visit because our curriculum best suited this visitor's expertise. Thinking of myself as a team player, I said I'd be happy to do whatever was needed to get my students ready for this visit. After all, I had great autonomy in my planning and could certainly tweak things for a command performance for one visitor. Then I asked the million-dollar question: Who was this visitor? To this day, I still recall my disbelief when the name Jane Goodall passed the headmaster's lips and hung in the air between us as if in slow motion.

After shaking off the amazement, I got to work figuring out how to weave Goodall's work into my curriculum. It soon became clear to me that she had a connection to the hominin skulls so prominently displayed near the science lecture hall. Leakey donated the cast skulls to St. Mark's; and Richard's father, Louis, hired Goodall to study the chimpanzees at Gombe in Tanzania. It made for a nice connection between school history, primatology, and human origins. So, inadvertently, my style of seeking interdisciplinary connections was born!

As I dove into the depths of Goodall's work with the Gombe chimps, I fell in love with the story again through the different lenses of a teacher rather than a young child. I was able to appreciate the study at Gombe as being clearly connected to a deeper study of human origins. This made me choose to teach my middle schoolers about Goodall's work not as merely a cute animal conservation story (the way I had been introduced to it in the 1970s), but as a way to compare and contrast human anatomy, social structure, and behavior with our nearest living relative. Once this connection was made for my students they became very engaged with a quest to better understand what that last common ancestor between chimps and humans might have been like.

The day Goodall visited St. Mark's was a triumph in my young career. Not only did my students and I get to meet and interact with a truly iconic scientist, but my students rose to the occasion by asking insightful questions that made me proud to have taught them.

I savored my time with Goodall, and her reply to my question of why she thought her work had become such a worldwide phenomenon changed my teaching forever. She explained to me that we as humans make deep connections through stories we share, and all she had done was share the story of her adventure and discoveries. She pointed out to me that children remember stories far better than lists of facts and in particular, they embrace the stories they hear at a young age. She had reached millions of young children with her story of the Gombe chimps and they responded to it just as children had responded to stories throughout human history. They listened to the story over and over and then shared it with friends and family members. She made it clear to me that the power of a good teacher (especially elementary and middle school teachers) was in our ability to connect young students with stories that would shape their understanding of the world long after leaving a particular classroom.

Goodall's take on the power of storytelling wound up profoundly impacting how I approached my teaching. I began to treat my teaching as in-depth storytelling, or, as I started to call it, "Sciencetelling." In particular, I began to shape my course into a curriculum that had a solid base in evolutionary stories—how we came to develop evolutionary theory (Darwin, Wallace, and their voyages of discovery) and how branches of the tree of life diversified and adapted over time.

This approach allows me to treat every scientific discovery as a story that connects science facts with the historical background of when and where it was made.

As my career grew, I was incredibly fortunate to have the chance to have several well-known evolutionary scientists visit St. Mark's: Richard Leakey's second visit to St. Mark's in 1994; Bob Bakker (*The Dinosaur Heresies*) in 1999; Louise Leakey in 2007; and Spencer Wells (*The Journey of Man: A Genetic Odyssey* and founder of National Geographic's Genographic Project) in 2009. I had great opportunities to build upon my base of science stories and tell new stories that related to real life scientists.

These great opportunities meshed with my love of human origins. My curriculum included an extended unit that covered primate evolution and Goodall's work; discovery stories of all the major hominin fossils; and

how those discoveries enhanced what we know of our own human story and what it means to be human. By 2010, I admit that I was comfortable with our understanding of the fossil history of humans. The fossil story had not changed dramatically since the discovery of *Ardipithecus* by Tim White and his team in the Afar region of Ethiopia in 1994. Like many others, I did not sense the field was going to change. Like many others, I was dead wrong!

During April 2010, two articles were published in the journal *Science* by Lee Berger describing a new species of two-million-year-old hominin from South Africa named *Australopithecus sediba*. Little did I realize how this discovery would change my life. Based on my background knowledge of human evolution, I was well-prepped to teach the science of *Australopithecus sediba* to my middle schoolers, but I did not have a grasp on its discovery story. As it serendipitously turned out, one of the scientists with whom I had become friends on Facebook was Berger, the discoverer (along with his then 9-year-old son, Matthew) of *A. sediba*.

One night in August 2012, as I was planning my curriculum for the year, I saw Berger's green Facebook light on and decided to message him to see if he would share *A. sediba*'s story with my students via email. I thought the worst thing that could happen was he could say "no" and then life would go on. Not only did he respond, but he agreed to participate and asked for some details about my school. When I told him my school's location, Berger mentioned that he was going to be in Dallas in November visiting friends on his National Geographic book tour promoting his new work, *The Skull in the Rock*, which detailed the discovery story of (you guessed it) *A. sediba*!

When I asked if he might be available for a morning visit with my students, I was again surprised by a positive answer. After more conversations and reaching out to our local science museum (the Perot Museum of Nature and Science), we added an evening talk that was open to the public. After a fantastic day with students and families, Berger shocked me by inviting me to his lab in South Africa to meet and study the fossils of *A. sediba* in June 2013!

I was completely floored by his generous offer . . . and then I proceeded to turn him down!

VIDEO RECOMMENDATIONS

The Origin of Species: The Beak of the Finch from HHMI BioInteractive is a great video. It details the work of iconic evolution researchers Peter and Rosemary Grant as they study the famous Galápagos finches on a single island (Daphne Major). The video helps connect the work of Darwin and his visit to the Galápagos with modern-day research. The couple's research demonstrates how environmental changes such as extreme drought and subsequent wet years changed the beak size of the resident finches. Their work also explores the mechanisms that help keep similar species apart. While not in the video, their more recent work shows these similar species have produced a new species.

I promise you that I was really not as crazy as you thought I was at the end of that last sentence. I had already signed a contract to teach a nature photography camp for children during the week Berger offered. However, in a "lemons-to-lemonade" moment, the camp changed its plans and a visit to Berger's lab became possible.

The week I spent in Johannesburg, South Africa, was an amateur paleoanthropologist's dream come true. I not only had the chance to see and photograph *A. sediba* fossils but also had the once-in-a-lifetime chance to hold the actual Taung Child fossil—the first early hominin fossil ever discovered in Africa.

My initial Facebook conversation with Berger not only allowed me to see some of the world's great fossils, but I found myself welcomed by the scientists at the University of the Witwatersrand. Despite not being a tenured university professor or even a PhD, I was treated like one as Berger and his colleagues entertained my questions and went out of their way to show me the nuts and bolts of how their lab worked. This even included a field trip to Malapa, the site where both *A. sediba* skeletons had been discovered.

I returned to Dallas on cloud nine and planned to weave the experience into my classes. Little did I know that on September 13, 2013, two South African cavers, Stephen Tucker and Rick Hunter, would discover hominin bones in the Rising Star Cave mere miles from Malapa.

This discovery caused Berger to launch a full-scale expedition into the bowels of Rising Star, where only the skinniest of scientists could fit. A public Facebook posting helped recruit a team of six highly qualified female scientists who became known as the "Underground Astronauts." They recovered more than 1,500 hominin fossils, more than any location in Africa. Breaking with the tradition of keeping a hominin fossil dig secretive, Berger and his colleagues began to live-tweet the hour-by-hour and minute-by-minute activity of the #RisingStarExpedition. My students and I followed it with great excitement, but I realized that much of what was being tweeted would be lost to cyberspace for those who did not use Twitter. At the happy expense of late nights, I began a daily video review of tweets from Rising Star so that they could be easily shared with students and other interested parties.

My "Twitter Play-by-Play" (http://bluelionphotos.blogspot.com/2013/11/rising-star-expedition.html) as it came to be known, allowed thousands of students and teachers to follow along from the start of the expedition even if they found out about it months later. It now serves as a primary source for interested parties wanting to dig deep.

As Berger's team worked to identify and study the fossils during 2014, I was able, thanks to social media, to build upon the relationships I had made online as well as in South Africa during my visit. This led to an in-person visit from underground astronaut Lindsay Hunter in January and a second visit from Berger in November 2014.

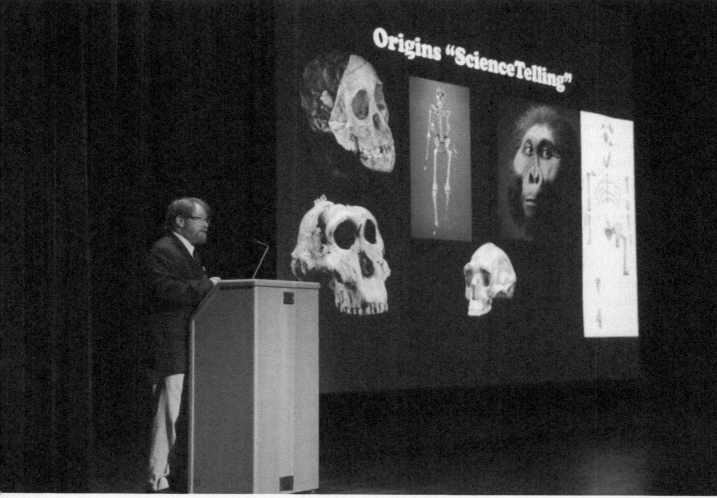

John Mead presenting about Lee Berger's discovery of *Homo naledi* to his students. Berger joined via livestream from South Africa.

Based on my work with the Twitter play-by-play project, I received an invitation to return to South Africa in July 2015 to visit the Rising Star Cave and see the as-yet-unnamed fossils in person. This time I did not refuse!

My 2015 visit mirrored my first visit. I was welcomed by the entire team as a team member myself. Indeed, they understood and appreciated that a classroom teacher could help the world understand this mysterious new hominin.

The highlight of the visit was having the chance to spend time at the cave and interview all members of the newly formed Hill Foundation Exploration team as well as the senior scientists. These interviews ranged from ten to thirty minutes and allowed the story of this discovery to be told to students and teachers in a depth difficult to find in mainstream media. Found on my blog (http://bluelionphotos.blogspot.com/2015/09/the-rising-star-interviews.html) and my YouTube channel, they are available for anyone eager to explore the Rising Star story in more detail.

After my two weeks in South Africa, I returned to Dallas sworn to secrecy. As a teacher who compulsively *shares* information, I found it torturous to keep the secret until September 10—the date of the official announcement.

BOOK RECOMMENDATIONS

The book I find myself most often recommending to fellow biology teachers (especially those without a deep background in evolution) is *Remarkable Creatures: Adventures in the Search for the Origin of Species* (2009) by Sean B. Carroll. I recommend this book for its superb writing and storytelling in addition to the breadth and depth of those stories. It covers the birth and formation of evolutionary thought before detailing seminal fossils discoveries from early tetrapods to dinosaurs to humans. Carroll does a masterful job of telling the adventure stories of discovery with the factual data that we need to have a deep understanding of various facets of evolution. It's a fantastic stepping stone for teachers to see evolution not as a simple theory to be taught but rather an idea that serves as the glue that connects and gives all biological facts relevance and meaning.

I had eagerly prepared my middle school students for this big news and was not worried that if I wanted to see it livestreamed I would have to be awake at 3:00 a.m. Dallas time. This would also mark the chance for me to *finally* let the world see my half-hour interview with Berger and John Hawks. I had bottled up all the details since my return in July.

I watched the official announcement livestreamed from the University of the Witwatersrand with much the same excitement I imagine accompanied the Apollo 11 moon landing in 1969. As Lee Berger's voice announced "a new species of human relative, *Homo naledi*" I made my interviews public and was stunned to see every news organization immediately had lead stories about *Homo naledi*. This was indeed a BIG DEAL!

I was thrilled that this fossil discovery, which I thought was important, was also being treated as important by the international mainstream press corps. But what followed was something that I did not anticipate in the least. If you recall, my initial connection with Berger was to have him play a small role in helping my students better understand his discovery of *Australopithecus sediba*. That was all I was hoping for back in that August evening in 2012.

I was unaware that in reaching out to Berger, I had done something rare for teachers, especially middle school teachers: I had networked beyond my close circle of colleagues. The fact that I had reached out to a scientist (admittedly one who was now in the spotlight) was seen as oddly newsworthy. In the immediate aftermath of the announcement, I was featured in the *Dallas Morning News, Scientific American*'s blog, the National Center for Science Education, and had a surprise visit from our local ABC news crew on a Friday afternoon.

Never in my wildest dreams could I have imagined that simply doing what was good for my students would have resulted in such coverage. Indeed, I was told on multiple occasions that I was "brave" for having reached out to Berger in the first place.

The more I thought about this situation, the more it bothered me that I was getting more attention than I felt was deserved for networking in a way that is considered the norm in almost every profession outside of teaching. I soon realized that such networking among teachers was indeed rare. Before internet availability, most teachers rarely had the time nor the professional connections to network beyond their school buildings.

Many teachers now successfully use social media to expand their personal learning networks, and I consider myself an eager proponent of this trend. We must encourage teachers in all disciplines to reach out to other teachers as well as experts in their fields so that students can benefit by gaining a wider and more realistic view of the world and how it works.

In the midst of the media excitement of the *Homo naledi* announcement, my students and I had the great good fortune to have Berger and three of the underground astronauts (Becca Peixotto, Hannah Morris, and Marina Elliott) visit Dallas to speak at the Perot Museum of Nature and Science. They were generous enough to spend time with my students and Berger stunned all of us when he presented me with casts of *Homo naledi*'s hand, foot, and skull. It was one of the few times in my teaching career I have been publically speechless!

As time moved on and things naturally quieted down, I settled into a routine of talking about the "*Homo naledi* Experience" to a range of interested groups. The story of the *Homo naledi* discovery is so engaging that all ages like hearing about it, and I have found myself educating thousands of interested folks from age five to 105 about this modern-day science/adventure story.

Much to my excitement, there remains no end in sight! Thanks to support from the Dallas-based Lyda Hill Foundation, the Rising Star Cave system has been purchased and the fossil site will remain protected in perpetuity. In addition, the Hill Foundation sponsors the team that explores the Cradle of Humankind looking for the next great hominin discovery.

Such support resulted in the May 2017 announcement of a second fossil chamber named Lesedi ("Light") within the Rising Star Cave system. The team announced the discovery of fossils of three individual specimens of *Homo naledi* in this area, including an almost complete skeleton named "Neo" (Meaning "gift") and fossils from an infant.

Upon the announcement of the *Homo naledi* discovery in 2015, one frustration for many people was the lack of a date associated with the bones. Based on the anatomy of the bones and how primitive some of them are, many people suspected that *Homo naledi* might date back to the dawn of the genus Homo about two million years ago.

After a lot of work attempting to date the fossils and the cave sediments, the team finally announced definitive dates for the Dinaledi chamber specimens in May 2017. Using several methods of dating (including electron spin resonance and uranium-thorium) processed independently by multiple labs, the bones were found to be from 236,000 to 335,000 years old.

This very young date surprised many people and suggests that *Homo naledi* was alive in southern Africa until the advent of our species, *Homo sapiens*, in the area. The potential presence of this small-brained relative alongside modern people has paleontologist Berger and his team questioning if *Homo naledi* might have been responsible for some of the stone tool artifacts found in South Africa at that time. This will certainly be the focus of future study in the archaeology of South Africa.

Many people were highly skeptical of the "ritualized body disposal" hypothesis since it simply seemed so implausible. Given the fact that all other reasonable causes for the presence of the bodies had been ruled out, the team still needed to work to disprove their own idea. If the only access to the Dinaledi chamber was the chute, it made sense that there should be remains of *Homo naledi* at the landing zone immediately below the chute. The 2013 fossils were excavated more than ten meters away from the chute.

In an attempt to gain evidence that might support or refute the chute-entry hypothesis, Berger reassembled most of the excavation team in September 2017 and began a three-week expedition to seek these answers, as well as to further excavate the Lesedi chamber hoping to recover more of the Neo skeleton.

The team discovered *Homo naledi* bones in a debris "cone" (pile) under the chute. There appears to be at least one skeleton in a mash of bones and sediments that will require further excavation. This discovery does offer strong support to the hypothesis that *Homo naledi* did indeed enter the Dinaledi chamber via the chute. To add new (and currently unresolved) intrigue to the chapter, new fossils were spotted in very narrow passages off the Dinaledi chamber—a discovery which will require further investigation. This last point gets to the main point of this chapter: As many students as possible should be studying the story of *Homo naledi,* and here's why:

The first and perhaps most obvious reason involves the great science this discovery has unleashed. As a self-confessed paleo geek, I think everyone should be well aware of the biggest human origins story since Lucy came on the scene in the mid-1970s. But the value of this story to young students goes well beyond the "bones and stones" of archaeology and paleontology.

As a teacher of middle school students, I love that this project has been active since 2013. I have former sixth grade students who are now following the latest

John Mead sharing hands-on resources with his students.

news as high school juniors. This is a great example that scientific discoveries are not just that magic *eureka* moment in a lab, but rather an ongoing series of hypotheses and investigations. It also shows students that answers do not come easily like the answers to questions at the end of a text book chapter or in a Google search.

In a world that values instant gratification, such a continuing investigation is a powerful model of how science really works. When you take the time to learn about the *Homo naledi* discovery, another thing that is important is the open access that Berger has made a hallmark of this project.

Unlike most discoveries in the field of paleoanthropology, each stage of the *Homo naledi* story has been deliberately open access. From live-tweeting all of the excavations since 2013 to wiring the Rising Star Cave with Wi-Fi so that students from all over the world could participate in hangouts with the team via the National Geographic Explorer Classroom program as well as through Facebook live broadcasts.

In addition to inviting students to visit the fossil chambers virtually, Berger's team has scanned many of the *Homo naledi* fossils and made them available to 3-D print for free from Morphosource.com. In my classroom,

I had 3-D prints of the original fossils within days of the official announcement. When holding precise 3-D prints of fossils that are in the news, students get especially excited and ask amazing questions.

As someone who has followed this investigation since Day One, I have gotten to walk my students through a great example of the scientific method in action. Rather than be a simplistic straight line from hypothesis to experiment to conclusion, the story of the *Homo naledi* discovery shows how science works in the modern world by combining observations, exploration, hypothesizing, and teambuilding.

Perhaps the most powerful of these characteristics is the impressive team nature of this project. Traditionally, human origins are studied by research groups of only a few senior scientists who have guarded data closely. As I've pointed out, the *Homo naledi* team was more than a handful of folks. The most visible members of the team were Lee Berger and Hawks along with the six underground astronauts and the two cavers who initially discovered the bones. However, what many people do not know is that the "team" consists of more than 150 scientists in the fields of paleoanthropology, osteology, geology, ecology, paleontology, geophysics, and exploration. Indeed, when you look at this team in comparison to other scientific endeavors, it is one of the largest science teams in any field and compares in size to the large teams that run NASA missions or make work at the CERN supercollider possible.

Another important "teamwork" lesson for students is the understanding that this discovery was made possible by a whole host of "nonscientists." For many students who do not see themselves as "good at science," this is an important point.

Homo naledi would not have become such a huge discovery if it was not for the following groups: The volunteer cavers who were responsible for the safety of the excavators, photographers, and videographers who documented the fossils and the work of the scientists; paleo-artists such as John Gurche, who created the iconic "grumpy" face of *Homo naledi*, tech gurus who helped scan the fossils and arranged for Wi-Fi in the cave so students would have real-time access to this discovery, camp management staff who kept the expeditions fed and running smoothly so that the science of this project could occur without interruption. You can also add teachers and science journalists who helped spread the news and significance of this find to people who otherwise would have missed it.

If what I've mentioned so far is not enough to convince you that the *Homo naledi* story should be widely taught, I also see it as a great example of what is a dying endeavor in the twenty-first century: *exploration*!

This scientific discovery has elements of exploration and adventure, which students often miss when they encounter science in their classes today. In our increasingly urbanized world, many people believe that "everything has been explored already." This project illustrates that even in the most heavily explored area on Earth for human fossils, there can still be a huge discovery. This lesson should be extrapolated for all of us to be thinking about what we might discover in our own neighborhoods that we think we know so well. If we take this lesson to heart, it can help us fight the narrow vision that comes with "Backyard Syndrome"—the tendency to overlook great things that exist right under our noses!

Since that day in August 2012 when I first made contact with Lee Berger, I've grown into what friends have called an "evolution evangelist" (a nice turn of a phrase, I think!). It initially started out with me sharing the *Homo naledi* story because it was interesting. By early 2016 and amid the excitement of the formal announcement and the *Homo naledi* story being named the second-biggest science story of the year by *Discover Magazine,* I assumed everyone was simply tired of hearing about *Homo naledi*. Then, on my way to a conference, I had the chance to meet a Facebook friend for the first time in person. He is a nationally recognized biology teacher who regularly works on evolution education issues. When he asked me to explain this "*Homo naledi* thing," I realized that rather than back off talking about it, I had to up my evolution education game.

I have had the privilege since then to speak to thousands of people about *Homo naledi* as a jumping off spot for learning about evolution in general and human evolution in particular. While I am always ready to encounter evolution deniers, this has rarely been the case. The primary reason for this, I believe, is my "Sciencetelling" approach which allows audience members to learn the scientific details from the story and make connections for themselves rather than being force-fed the "right" answers. I also make certain to explain how this science adventure does not pretend to have all the answers and that the tools of science investigation allow us to continue to seek better understanding of our current gaps of knowledge. My way of teaching evolution by entwining it in the larger web of adventure and discovery and using it as a way to also teach the nature of science allows students and those uneasy with the concept to "come along for the ride" and have the chance to have their eyes opened in a nonconfrontational way.

When presented with a story of genuine discovery, most people welcome the information and follow the line of logical reasoning because they have not been battered over the head with dogma or been told what to believe. Rather they appreciate that they are respected enough to come to their own conclusions and are free to ask genuine and occasionally challenging questions. So much of the evolution/creation battle has been focused on the extremes telling the middle ground what to "believe" that we have lost most of that middle ground because of simple avoidance of the whole discussion. When we as evolution educators approach teaching evolution as a chance to have meaningful conversations with people who are often uncomfortable (but not hostile) to the topic, we have the chance to make great inroads into an acceptance of both evolutionary theory in particular and scientific understanding in general. As the number of evolution educators in the elementary and middle grades grows, it expands the exposure of children to a topic that most find highly sensible. Given that the vast majority of Western children grow up with regular exposure to Judeo-Christian origins stories through either their faith tradition or general Western culture but minimal exposure to the basics of evolutionary thought, it should be no surprise that societal understanding and acceptance of evolution are lacking. With these ideas as my guideposts, I plan to continue to move forward seeking to help fellow teachers in both formal and informal settings become passionate advocates for improving both the quality and the quantity of evolution education worldwide. I am honored and excited to be a part of the TIES community.

In my years being connected with TIES, I feel it has done more to reach both teachers and students who are new to evolution than any other evolution education group. TIES looks not to preach to the choir, but rather seeks to expand the tent of people who can appreciate the beauty and wonder of evolution. I am optimistic about what the future holds thanks to my wonderful TIES colleagues who have brought TIES workshops to more than forty-two states in person and to a wealth of others though online webinars and social media outreach. By building a community of evolution educators through TIES and other evolution-centric organizations, the future of evolution education and acceptance shines brighter than ever!

Introducing Evolution

 My favorite bell-ringer for my sixth graders demonstrates to students that natural selection operates on the interplay of variation in a population and the change in environment over time. It works by each student choosing a card from a deck. Each card starts with a "fitness score" of ten points. For each "generation" a random generator I programmed chooses a group of cards (all the face cards, all red cards, all the sevens, all the clubs, etc.) and then also chooses a "fitness adjustment" from -10 to 10 for that group of cards for that generation. This fitness adjustment broadly represents climate and its impact on our card population. We run the simulation for a set number of generations to see which groups of cards are most "successful" and which cards go extinct. The population of cards has fifty-two variations which serves as a model of genetic variation and the random changes in the fitness adjustments model environmental changes and impacts that can be positive, negative, or neutral for any of the fifty-two card variants. By doing this over the span of many days, students see that while variation and environment are both important, evolutionary success comes from the interplay of the two. It also teaches the value of genetic diversity as they soon see that a homogeneous population can easily go extinct. This bell-ringer can be expanded into a longer full period activity where groups are divided into separate populations with differing climates.

Favorite Investigative Unit or Activity

There are so many great evolution labs out there but with the advent of 3-D printing and increased availability of hominin skulls, my favorite lab is to do a hominin craniometry lab that allows students to handle 3-D prints of skulls and measure cranial capacity, facial angle, tooth sizes, and foramen magnum position across a range of species (I use *A. afarensis, P. boisei, A. sediba, A. africanus, H. erectus, H. habilis, H. neanderthalensis, H. naledi, and H. sapiens.*) printed out from the Human Evolution Teaching Material Project website: https://www.hetmp. com/ For teachers without access to 3-D prints, the original version of this lab based on photographs is available on the TIES website.

Dealing with Conflicts

In my thirty years of teaching evolution, I am fortunate to have had relatively few negative experiences with students pushing back on the validity of evolution. However, when I do have pushback, there are two things I use that help to defuse vehement opposition. Given that I teach middle school students, my focus on teaching the exploration and discovery stories of human evolution take lessons from being just about "evolution" per se and makes them more relatable as adventures in science. This also lessens the misconception that scientists are trying to prove that "God does not exist" or that "Religion is wrong." Teaching from this discovery perspective also lets me focus on the nature of science and lets students see that the evidence stands on its own.

Occasionally I get a student who is anti-evolution and has a strong contrarian personality. These students are eager to make a show of proving things wrong and they often like to focus on evolution as a target. Given that my students are young, these particular students often have poorly developed debating skills and I am usually able to convince them that to attack evolution successfully they must first understand it. I then offer to give them tutoring sessions outside of class so they can sharpen their debate skills. These students like that I care enough about them to develop their skills, and it winds up being a positive personal relationship rather than a combative one. In addition, I remove the conflict from class and have time to show them why they are wrong in a detailed, respectful, and supportive manner. In each of the cases where I have implemented this strategy, I wound up with a stronger relationship with the student as well as a much more positive classroom atmosphere.

Chipmunks, Eternal Damnation, and Other Hazards of Biology

David Mowry, BS in Wildlife Biology
Bremen, OH

Athens, Ohio
June, 2004

Despite holding multiple doctorates and undoubtedly decades of education between them, apparently none of the faculty occupying the graduation stage had considered that giving a lawyer an honorary PhD and free rein on the microphone was a rather questionable idea.

I sat between my roommates and friends, waiting to be handed my new degree in wildlife biology and, while doing my best to ignore the droning lawyer, wondered, "What do I do next?" (I would learn later in life, when I began to pursue education as a trade, that this process is called "reflection.") I had been offered a job as a field technician by one of my professors conducting a mark-recapture study, which would last through the end of the year. I had always been inclined towards the sciences, particularly biology, and this seemed like a natural fit. A few weeks later, I eagerly began my new job.

As it turned out, being a field technician involves, in my case at least, running around in the woods by yourself for hours and getting bitten by chipmunks. Much to my surprise, while I was (and remain) grateful to my professor for giving me the experience, I decided that this area of the sciences wasn't for me. I missed interacting with people which, always having been introverted, also surprised me.

When my contract was up, I began exploring other options. My mother ran across an advertisement for an outdoor education position at a camp a few hours away from where I lived. I sent in a résumé just because

I needed interview experience. I had no real intention of actually taking a teaching job, because of the aforementioned introversion. Much to my surprise, I received a phone call a few weeks later informing me that I was now a teacher.

I arrived at Nature's Classroom with no education background and no education experience at all. I was, however, one of the few staff members with a background in science. Oddly enough, it seemed that people with degrees in science rarely applied to this type of job. After two weeks of in-service and one long weekend of trying to figure out how to make lesson plans, we were swarmed by one hundred or so sixth graders. I had to keep them safe and happy (or at least alive) for an hour and a half and hopefully teach them something about nature along the way.

It was a fantastic job. There were some bumps on the road, like with any job, but for the first time in my life, I felt like I really had a clear direction to pursue. I realized that, in my job as a field technician, what I was really missing was sharing things that I thought were interesting with other people. As one of the few staff members with a biology background, I frequently found myself teaching the more biology-oriented classes, such as our herpetology class, geology and natural history, stream exploration, and many others. The topic of evolution even came up from time to time, even though it's not something that, in my experience, is typically addressed in a middle school setting.

Sometime during my second season at Nature's Classroom, we were told that a private religious school from my home county would be coming in. Other private schools attended from time to time, so I didn't think too much of it. There would be grace before meals and some morning devotions, but nothing that would really affect my routine. One of my cousins had attended this school earlier, and I wasn't really aware of it beyond that.

The three days passed largely without incident until Friday morning, as we were preparing for our last hike. One of the topics that we covered on these hikes was adaptations, including some that humans have. Our director was gone that morning for a school meeting, leaving our assistant director in charge. At our daily after-breakfast meeting, as he was laying out the plan and jobs for the day, he suddenly looked up and said, "Excuse me, I need to check on something." He got up and wandered off. The rest of us, thinking this was unusual, but not having any other instructions, went about our normal Friday checkout routines.

An hour or so later, after cabins had been cleaned and luggage packed, the staff met on the basketball court where our hikes began. The students were still finishing their morning devotions and hadn't arrived yet. The assistant director pulled us all together and said, "OK gang, I've checked with the lead teacher and it's OK for us to call humans animals and mammals this year."

Before I could stop myself (my introversion was gone by this point, replaced by a general policy of, "say it anyway, and see what happens") I heard myself asking, "What else could we call them?"

I hadn't realized exactly what type of religious school we were dealing with, up to this point.

I am not antireligious by any means. I was raised in, and remain a member of the Presbyterian tradition. My parents took us to church every Sunday, and my father was a university biology professor. It was made very clear to me early in life that science and religion do not need to be in conflict with each other. The Presbyterian tradition has largely been very supportive of science, including evolution. It was simply a nonissue. I had dealt with fundamentalists before, and, to my discredit, probably not as gracefully as I could have.

Yet here I was, at a camp where kids had come to learn about nature, and we were debating how to pussyfoot around a central tenet of nature and try not to alienate a paying customer to boot. It was an interesting situation, to say the least, and no real adequate solution seemed to present itself.

Before we could go any further on the subject, the kids arrived from their morning devotions. As we broke into groups and went our separate routes on the hikes, I noticed that some of mine were crying. Kids at camp should never be crying.

My family generally deals with feelings like door-to-door salesmen: ignore long enough, and they'll go away. I was a little out of my depth here with crying children, but I inquired as gently as I could. They told me that during their morning devotions, they had discussed various things that could lead to eternal damnation (which seems like a super-great way for a 12-year-old to start the day). One of the things on the list was evolution. A couple more questions revealed, among other things, the usual misconceptions about the idea (man came from monkeys, etc. . . .). Some of these students had been frightened into tears because of a subject that they barely understood. Now, I was angry and determined. I didn't want to see another child ever have to cry because they were frightened of something they didn't understand. I finally had my answer to that question I had asked myself a couple of years before during graduation.

Challenges of Teaching Evolution in the US

This was more than a decade ago. My memory has faded, surely, but those are the details as best as I remember them. And the determination that began that day has guided everything that I have done since. I left Nature's Classroom behind and went to graduate school for science education. I had decided that the best way for me to make a difference was to become a high school biology teacher. I made a promise to myself that my students would understand evolution, how it works, and why it's important before I was done with them.

After I left Nature's Classroom and completed another year of undergrad and a couple of years of grad school, I found myself and my brand new teaching certificate employed at a career-technical school. In

VIDEO RECOMMENDATIONS

My favorite video is *The Origin of Species: The Making of a Theory*, which can be found on the BioInteractive program from the Howard Hughes Medical Institute. This film details the journeys of both Charles Darwin and Alfred Russel Wallace and how they independently developed their theories of natural selection. It runs about thirty minutes, which is basically an entire class period at my school, but I've found it to be thirty very well-spent minutes.

I like this video for a variety of reasons. It's an excellent outline of how we arrived at one of the most important ideas of modern biology and the impact that Darwin and Wallace had. It puts the development of their ideas into historical context for the students, which really helps them understand, but all too often probably gets glossed over. It also deals very nicely with the evidence that they used to support their claims. It shows the students that this is not some unsupported, speculative claim, but a well-thought-out theory, with a mountain of evidence, collected by different people all over the world, pointing to the same conclusion. HHMI also provides several worksheets to accompany the video. It does touch on how some of Darwin's and Wallace's ideas conflicted with established religious beliefs, but does so in a way that has never created arguments, at least in my classes.

southeastern Ohio, evolution can still be a tricky subject, but I intended to stick to my promise of making sure that my students understood how important it was.

I spoke with some of my new colleagues and they told me that the fight over whether or not to include it in the curriculum had already happened years earlier in the district. One, who had taught biology in the past, said that she avoided the topic altogether because it wasn't worth the hassle. The state science standards had changed since then, though, and it wasn't really optional for me.

Then, one of our career-technical instructors, with whom I work closely, said to me, "It's very important to pick your battles. Of course I was married, so I pick all of them." I've never forgotten that piece of wisdom. I would teach evolution and teach it well and dig my heels in and stand my ground when the time came.

As it turned out, the knock-down, drag-out fights I was expecting never really came. Most of the students, if they had any issues, largely kept them to themselves. I only really had one problem with a student, who said he opposed evolution on religious grounds. His religious convictions really seemed to falter when I told him he could believe whatever he wanted, as long as he passed

the test. I'm not sure if this was the best approach, but it did stop future arguments.

One of the things that quickly became apparent was that most of the students' misconceptions about evolution come from not understanding the nature of science. This has been discussed elsewhere in this book as well. In Ohio, like most of the US I suspect, although admittedly I haven't looked it up, standardized testing is heavily emphasized. Our content standards are generally several paragraphs and bullet-point lists that mention vague scientific topics, and we have to magically divine out of this what it is the state would like our students to know for the test. As a result of this, most of our time is spent cramming as many facts as we can into our students' heads in the hope that some of them might be regurgitated on a multiple-choice question a few months later. Scientific methodology is mentioned as something that should be taught throughout the course, but in practicality, I doubt this ever happens to the extent that it should.

One of the first things that I do at the beginning of each class is to try to clear up many of the misconceptions about science and the scientific method. Materials from the Understanding Science website from the University of California Museum of Paleontology have been an invaluable resource for this (Understanding Science 2018). When time has permitted, which is unfortunately becoming less and less frequently, I will use the website as a yearlong project, having the students work through each section as we work our way through the regular course content.

Another thing that we do is to spend a bit of time on the difference between a hypothesis, a theory, and a law. I doubt that I'm alone, but I have noticed, year after year, that students think the three are interchangeable terms.

Again, the "Science at Multiple Levels" page from the University of California has been an invaluable resource in this. A quick Google search reveals many examples and worksheets that students can use. The example that best gets the point across is attributed to TIES esteemed patron Professor Richard Dawkins, who is quoted as saying, "Evolution is just a theory? Well, so is gravity and I don't see you jumping out of buildings." I don't immediately recall which of Dawkins's works this comes from, but it really seems to drive the point home to the students about our level of certainty regarding evolution.

I spend an entire quarter on evolution. I certainly don't ignore it during the rest of the year, and we discuss it as we work our way through cells and genetics and our other topics, but I want to make sure that students understand its importance to biology. Also, it's a complex topic and, like anything else, if you're going to do it, take the time and do it right.

A couple of years after I started teaching, my wife and I attended a workshop put on by the Smithsonian Institution's Human Origins Program (2018). Connie Bertka, of the Broader Social Impacts Committee, briefed us on some of the social and cultural issues facing evolution educators today. Briana Pobiner, a paleoanthropologist, taught us how to correctly complete craniometrics on (replica) hominid skulls. It was an invaluable piece of professional development (and a rare one as well, since Ohio doesn't require professional development in a teacher's actual content area). I noticed, though, that most of the people who attended were surprised when I

said that I spent an entire quarter on evolution. In turn, I was surprised when they told me that many teachers spend significantly less time than that on it, choosing the path of least resistance. This reinforced my decision to teach evolution and take the time to do it right. Even though Ohio's end-of-course exam specifically states not to assess human evolution, I acquired a set of replica hominid skulls and Pobiner's skull craniometrics activities has become one of my students' favorites. (The end-of-course test prep also tells me to teach them the Hardy-Weinberg equations, without teaching them how to calculate Hardy-Weinberg equilibrium, so I kind of feel stuck between a rock and a hard place often.)

As we work through evolution, I also use the Understanding Evolution resources, also provided by the University of California Museum of Paleontology (2018). Like its other website, a fantastic wealth of resources is provided here. I have found two of them, "The History of Evolutionary Thought" (2018) and "Evolution 101" (2018), to be particularly helpful to my students. For my students, I formatted information from both of these into webquests

and have them explore each site for the answers. I have found that this really drives home our certainty about evolution. The history section reinforces that this is an idea that has been developed over time, by some of the best and brightest that our species has to offer, not just an off-the-cuff thought by Darwin, as I have noticed it is sometimes presented by critics. "Evolution 101" provides an invaluable overview of how evolution works and the overwhelming amount of evidence that we have backing it up.

Another resource that has been invaluable to me are the materials available by the BioInteractive program of the Howard Hughes Medical Institute (HHMI 2018). I am particularly partial to the short film, *The Making of Theory,* which details Darwin's and Wallace's journey's as they arrived at the idea of natural selection independently. *Fossils, Genes, and Mousetraps,* a video featuring Ken Miller, also effectively explains why evolution is true. I don't use this one directly in class, but it has

BOOK RECOMMENDATIONS

Evolution vs. Creationism: Inside the Controversy, (2017) published by the editors of *Scientific American.* doesn't deal with how evolution works specifically, but provides an overview of the evolution/creationist controversy in the US. It is composed of a series of essays broken into four sections. The first deals with Darwin's legacy and the evidence for evolution that includes an essay by Peter Grant regarding his work in the Galápagos with Darwin's finches. The second and third sections cover the history of creationism in the US and how it affects today's classrooms. The fourth section discusses the broader topic of faith and science and includes a contribution by Richard Dawkins and Lawrence Krauss. The final section addressed America's "science problem" and the future.

Although it's more about controversy than evolution in particular, this book is an invaluable resource. Background and context are important to dealing with any problem, but especially one as old and complex as this. It gives an excellent background on creationists, their beliefs, and how to respond to them. It also gives a nice overview of what is and isn't permissible in American classrooms. I can't recommend this book highly enough to people who live in an area where there is social (not scientific) controversy on this issue.

provided enormous help in dealing with student objections to evolution.

I could probably go on for pages about the wonderful activities regarding evolution that are available now, but I think you probably get the idea. These have helped me greatly over the years, and I continue to find new resources all the time.

Since I started focusing on the nature of science and how it works and since I devote so much time to and assign so much importance to evolution, student objections have lessened over the years. This is an encouraging sign to me, and I'm hoping the trend continues, but it seems to me that, by and large, public opinion is not necessarily shifting in an encouraging direction.

I would like to offer a word of encouragement. Facing opposition to evolution is not going to go away anytime soon. Unfortunately, that seems to be a part of the job for us biology teachers. Still, we have several billion years of evidence on our side, which is a pretty good ally to have.

As I'm finishing this up, my 4-year-old daughter, Ezri, is belting out "Thank God the Tiki Bar is Open,"

at the top of her lungs, and my 1-year-old son, Dipper, is trying desperately to "help" me type this. My own two personal pieces of evidence for evolution. "There is grandeur in this view of life," Darwin famously stated. Looking at my children, I am inclined to agree with him. I don't want them to grow up in a world where a basic process of science is misunderstood and ignored. Evolution is crucial and amazing and almost magical to me. (That's OK, though. Remember that Terry Pratchett said, "It doesn't stop being magic just because you know how it works.") We have a responsibility to share one of the most amazing natural processes with the younger generations, so that they can, and will, accomplish and learn things we haven't even thought of. Don't be shy. Don't give up. And, in the words of my colleague, when the time comes to pick your battles, pick all of them.

Introducing Evolution

Since so many students don't understand how important to biology the concept of evolution really is, or are just flat-out misinformed about it, my approach to introducing it is like my dad's approach to voting: do it early and often natural selection, to name just a few examples). Furthermore, I don't debate the fact that evolution is real and happening. There's no debate in the scientific community and I don't like to give the illusion that there is, since this is an area that I've noticed students tend to be particularly misinformed about.

When I come to the formal evolution unit, about half-way through the year, I like to introduce it with an activity called Modeling Evolution: The Charlie Shuffle. This activity was developed by Vanessa Jason on Teachers Pay Teachers and is one of my favorites.

The activity works by giving each student a set of 4 cards. The cards say, "live," "die," "reproduce" or "mutate" on them. The students will then shuffle their cards and move around the room to exchange cards with their classmates. The goal is to, over the course of multiple rounds, survive and reproduce. If they get two "die" cards in any of the rounds, they get picked off by natural selection and die. If they get two "mutate" cards, they have mutated and are too genetically different to be considered members of the original species and are out. When the activity is finished, there are some wrap-up questions that accompany it and we discuss why some of the students survived and why some did not.

Keeping costs under control is important, and even something simple such as children's toys can be used to discuss evolutionary relationships over time.

(or so he says). At the very beginning of the year, in the first chapter of our textbook (Savvas Learning Company's *Miller and Levine Biology*, by Ken Miller and Joe Levine), we discuss evolution as one of the unifying concepts of biology and how it is responsible for both the unity and the diversity of life on Earth. We emphasize that, without it, biology would essentially be nothing more than a set of flashcards with neat, but unconnected facts about plants, animals, or whatever on them. As the year goes on, we talk about evolution when it comes up in the curriculum (how mitochondria and chloroplasts evolved, how mutations and genetic variation can drive evolution and

I like this activity for a variety of reasons. It is easy to do and it really addresses the main points of evolution in a way that is easy for the students to understand, even if they don't know all the terms and technical details yet. It shows the random nature of the effects of mutations and the non-random nature of natural selection preserving beneficial traits and weeding out less desirable ones. Even if the students don't put all the pieces together immediately, later in the unit when we come to these topics, they tend to remember the results of the game and then everything falls into place.

Additionally, this activity introduces evolution in a way that makes students comfortable with it. I live and teach in an area where religious opposition to evolution is still a concern. It has already been noted that evolutionary biologists don't really distinguish between micro- and macroevolution. This lets the students see that evolution can occur on a small scale, microevolution, which is almost impossible to deny. Even the most hard-core creationist would have a hard time arguing against something like this. By the time we work our way up to the process of speciation, the students generally understand that there really isn't any difference between the processes. Using this approach, I've found it's generally a pretty effective way to introduce the concept of evolution and to head off potential conflicts down the road.

Favorite Investigative Unit or Activity

I use the Skull Craniometrics activity from the Smithsonian Institution's Human Origins Program. I use a set of hominid skulls and have the students take cranium measurements.

They use string, rulers, and graph paper to calculate brain case size, face slope and brow ridge size on each of the skulls. I use replica skulls for *Australopithecus afarensis*, *Australopithecus africanus*, *Homo habilis*, *Homo erectus*, *Homo neanderthalensis,* and *Homo heidelbergensis*. When the measurements are taken, we then will graph out the results for all three traits and compare them. If they are done correctly, the face slope and brow ridge sizes should shrink and the braincase size should increase as they move from the older to the more recent skulls.

One of the biggest reasons I like this activity is that the students can measure skulls just like paleoanthropologists in the field. They can also see for themselves evidence of change over time and it's a great illustration of the diversity and unity that evolution produces. They see with their own eyes a group of different, but obviously related, organisms. Since they can tell that they are related to us, it really tends to hit home and be relevant to them.

Opposition to evolution is a concern, since this activity blatantly deals with humans. I do this at the end of the unit and by this time we've spent most of a nine-week grading period on the topic. At that point, most of the students have a pretty good idea of how evolution works so I really haven't encountered any resistance to it and they really seem to enjoy it. The vast majority of the students who have done this are fascinated by getting to investigate our ancestors. Even so, I've had such good luck with this activity, if I had to pick just one, this would be it, even if I knew I was going to have resistance to it.

Dealing with Conflicts

The only major conflict that I've ever really had to deal with was early in my career, when a student announced to the class that he didn't believe in evolution. He'd given no indication of any problems before this, so I was a little surprised. He never really said why (and I didn't ask), but I can only assume his rejection stemmed mostly from his religious beliefs and partly from his pathological need to be the center of attention.

I didn't want to get into an argument with him for several reasons. First, his personal religious convictions, whatever they may have been, don't overturn billions of years of evidence no matter how passionate he is about them. I make it very clear to my students early in the course that science depends on evidence. Personal feelings, opinions, and beliefs shouldn't factor in or should at least be minimized as much as possible. I don't even usually mention evolution while we are discussing this part of the nature of science, so by the time we get to evolution they know that the evidence says what it says and that's it. Whether you accept it or not is your problem, but billions of years of facts don't get thrown out because you've decided to disagree. I tell them that I personally would find the sky prettier if it were green instead of blue, but every morning when I wake up, all the evidence says that the sky is still blue, whether I like it or not.

Second, I am not a legal expert by any means, but I think that discussing creationist ideas in a classroom is at least definitely bending the rules about the separation of church and state here in the U.S. On a purely cynical level, that's just one more headache I don't want to deal with.

Finally, when I was younger, my parents both said that we shouldn't argue with creationists. Not that we should back down and not stand up for our positions, but because arguing with them gives the illusion that their ideas have scientific merit, when they most certainly do not have any. It would give the impression that creationist ideas are scientific and have an equal footing with evolution. An alternative, as it were. If a person is a creationist, there's probably not much I can do to change his or her mind, for better or worse. But I'm not going to argue and debate and treat a religious position as a valid scientific viewpoint.

I finally told that student that I didn't care what he believed personally, as long as he passed the test. I don't know if that was the best I could have done or not, but he got an A on the test, so I hope he at least understood it, even if he didn't buy in.

11 The Nature of Science, Evolution, and Storytelling

Blake Touchet, EdD
Abbeville, LA

As a child with boundless curiosity and a huge imagination, growing up in a tiny rural town in South Louisiana was always an adventure. Looking back, it seems almost unavoidable that my first two passions were exploring and reading. I must have investigated every square foot of every acre on our homestead for frogs and plants and interesting bones. By the time I had started kindergarten, I knew which insects and grasses would sting if I tried to play with them, where to find all of the honeysuckle and blackberries, and I had discovered a collection of skulls from animals around our property including sheep, horses, cattle, cats, and even a coyote. I used to make up stories about where they might have come from and what kinds of lives they may have lived. I spent countless hours observing the behavior of chickens and ants, watched the births of rabbits, pigs, and calves, and sat fascinated as spiders spun their intricate webs. And if there was ever a curiosity that I could not find the answer to on my own, I knew that books would hold the answer.

At first, my fascination with books was purely scientific. I needed answers, and books were sources of information. I wanted knowledge, and books were the keys to unlocking it. Later, after I discovered the joys of science fiction and fantasy novels, I read as a coping mechanism. When the tangible world around me became too overwhelming or intense, I could hide away in Narnia or Middle-Earth until the problems facing me seemed far away and insignificant. I am sure that my dad thought I was the strangest child he had ever known.

Upon arriving home from school, I would promptly go outside to find a place to do my homework where I was most comfortable, which was usually under a tree or up on the roof of our shed. When my homework was done and the sun was going down, I would shut myself in my room and become immersed in a different world. My dad always came to find me when he got home from work to ask me what was new. Looking back, I can only admire the love and patience it must have taken for him to try to follow my conversations, which jumped from the hard reality of science to the fantasy realms in my stories, but never anywhere in between. "Look at this neat slug I found on the driveway; doesn't it remind you of the time Ron's curse on Draco backfired and he vomited slugs all day?" But he only ever had words of encouragement and would sit in my room with me so that I could tell him all about it.

These dual passions of science and literature taught me many things growing up. From my observations and investigations, I learned that everything in nature has a logical explanation even if it is not easy to understand or immediately obvious, and that sometimes natural explanation is more elegant and wonderful than anything anyone could have imagined. From the stories in my books, I learned empathy and courage, how to see the world through the eyes of another, and how to persevere even when life seems difficult. These two passions still hold fast in my mind and personality to this day and have tremendously influenced my life and career path.

When contemplating career choices so that I could decide upon a college major, I thought that my ultimate goal was to become a research scientist. I imagined that I could be happy working in wildlife conservation or agronomy. I soon discovered, however, that I could never be satisfied unless both aspects of my personality were actualized. So what better way for me to share my love of science and literature with the world than to major in secondary biology education and minor in library science? My dad now had the rest of the world for company in thinking I was crazy, but I was happy as a clam in an intertidal zone. I learned to love the feeling of sharing in discovery with students and seeing the same look of excitement for science from them that I see in myself. I also discovered that showing students how to see the world from a scientific perspective allowed them to understand the beauty and interconnectedness of all living things in a way that enhanced the fun and wonder of life rather than diminishing it.

After I had been teaching for a few years, I decided to go back to school for a master's degree in biology. This decision stemmed from the fact that students sometimes had questions that I did not know how to explain, and I thought that more content knowledge would help me become a better teacher. However, the most important thing that I discovered, aside from the fact that I never want to stop learning, was that having more knowledge about biology helped me answer some of their questions, but not all, and not the important ones. The easy questions are the ones that teachers do not necessarily need content knowledge to answer. A simple Google search will suffice to satisfy their curiosity, and it is good for students to see that even teachers do not know everything. In fact, searching for answers with students can lead to some wonderful in-class investigations. The hard questions are the ones that cannot be satisfactorily answered even by the most knowledgeable professional. The students had doubts, questions of "Why?" and questions of "What if?" These are not questions of content. They are questions of process and philosophy and epistemology. The students want what we all wanted when we were children. They want to be reassured that what they are being taught is not only real, but also important and safe

and that it can be incorporated into their own understanding of the world. They want to know what is true, how we know that it is true, and how to deal with a truth that conflicts with their prior knowledge and beliefs.

It is fair to say that all science teachers have met with at least some resistance from students when it comes to teaching topics that the general public deem "controversial." These include subjects that evoke deeply personal and emotional responses such as evolution, climate change, GMOs, and vaccines. However, while these topics may be controversial to some in the general population, they are not controversial to scientists (Funk, 2015). If there is a disagreement between scientists about any of these ideas, it is not a dispute over the efficacy or legitimacy of the entire concept. Rather, experts may squabble over minor details of the interpretation of a very specific data set. For instance, biologists may disagree on what had a greater impact on the speciation of a particular beetle in a pine forest: was it natural selection because of changes in annual rainfall or genetic drift caused by a 400-acre fire a decade ago? Neither camp doubts that evolution has occurred or even what the results are; they simply disagree on the extent to which different mechanisms played a role in that process. Climatologists do not deny the fact that carbon dioxide and methane are greenhouse gases, that humans are a major contributor of these gases, or even that these contributions are leading to warmer surface temperatures. The disagreements, again, may be in minor, specific details such as what the precise rate of warming is, what percentage of these emissions are being absorbed by the ocean rather than the atmosphere, or how these variables will impact specific, diverse locations on different scales.

Scientists understand these subtleties, but the question is how do we get students and the general public to understand and accept them? The answer, of course, is education. But what do we teach? And when do we begin? Obviously, there is not enough time, even in an entire K-12 education, to teach every student every intricate detail of every topic. Indeed, people devote their entire lives and careers to the study of science and may only succeed in making progress in one small area of a single field. So are we to abandon hope of ever having an educated, informed citizenry? Of course not! The solution, then, is to decide how best to teach students so

VIDEO RECOMMENDATIONS

When agonizing over the attempt to narrow down my favorite short videos on evolution, my only consolation was that the other teacher-authors in this book would include all of the ones that I could not. Even with that comforting thought in mind, it still took many sleepless nights, dozens of re-watches of all of my favorites, and a poll from my students to finally decide on a single video. It was the student poll winner that I have chosen to discuss here. As always, my students impressed me with their insightfulness to see the bigger picture.

One of my favorite video explanations for evolution comes from the PBS Eons team. Although every one of their videos is amazing, the one that does the best job of fully explaining the history, processes, and intricacies of evolution is the episode titled "How Evolution Works (and How We Figured It Out)" from season 2 (2019). This twelve-minute video starts with a quick overview of the theory of evolution by means of natural selection and some of the early ideas that influenced Darwin and Wallace in their development of the idea, including the work of Lyell, Lamarck, Malthus, and others. After demonstrating the basic concept of natural selection, the video goes on to discuss the scientists who contributed to the development of the modern synthesis after Darwin and Wallace such as Morgan, Wright, Fisher, and Haldane.

This video clarifies a number of misconceptions about both evolution and the nature of science. First, many students have never heard of any evolutionary biologists besides Charles Darwin, and many hold the misconception that Darwin came up with the idea of evolution all by himself and that his word was the only and final say on the topic. This video shows that the development of our understanding of evolution and how it works has progressed in small incremental steps beginning well before Darwin and continuing up to the present day. This is a great demonstration of the process of science. The Eons team pieces this story together in an interesting and coherent way to show students how all of these discoveries build upon one another and add to our continually growing understanding of how populations change over time.

The video also details additional mechanisms of evolution, including mutations, genetic drift, and gene flow. It also discusses how they were discovered and what their influences are on the process of evolution. For my students to have voted this video as their favorite makes me very proud. It shows me that they understand the bigger picture. They see that not only is knowing about the process of evolution important, but recognizing how the process of science works is also important. My students were able to recognize that scientists never stop questioning and learning. Although Darwin didn't know everything about how evolution worked, that doesn't mean he was totally wrong. It also shows that science is more often a long, slow collaborative journey and not a solitary "aha" moment. My students liked how scientists were able to take two separate ideas from the nineteenth century (natural selection and genetics), put them together, and then build upon this synthesis to bring our knowledge and understanding into the twentieth century and beyond. They recognize that this has to happen continuously if we are going to keep advancing forward and unraveling the mysteries of nature.

that they can learn the basic foundations of these topics while, most importantly, gaining an understanding of the nature of science itself. Only by fully understanding the process of science can we expect people to place their trust in the data that have been collected and the conclusions that have been reached by its analysis (Lederman, 1992). Students should be introduced to the practices and processes of science early and be provided opportunities to practice these throughout their education careers.

In my experience, the area that leads to the majority of the issues when it comes to not accepting a scientific idea is an incomplete understanding of the very nature of science itself. If we teach students this concept early and encourage them to develop a full comprehension of it, then they will be better positioned to view everything through the lens of scientific understanding. By looking at the world through this filter, they will be able to judge the merits of information for themselves in order to make decisions (Musante, 2005). Our goal as science educators is not to radically change a child's worldview or self-identity, but to foster an attitude of open-minded curiosity tempered with rational skepticism that they can apply to scientific topics in the classroom. Ideally, this will also lead to the use of these skills in intelligent voting decisions, wise health choices, economic responsibility, and even social justice advocacy later in life.

Many excellent resources exist for teaching the nature of science and connecting it to specific content, especially evolutionary theory (Evolution 2001; Resources for Teaching and Learning Biology 2018). To summarize the major concepts of these resources however, science can be defined in three distinct, yet interconnected, ways. 1) What we know. 2) How we know what we know. 3) How we know what we know is actually true. The first definition of science would function as a noun. This is probably the definition that most people think of when considering the word. Science, according to this use, is all of the collective information that we have discovered about the universe. Although this is not an incorrect definition of the word, many people do not go past this simple application. This is where part of the problem arises, as science is so much more than just a random assortment of trivial facts. It is important to not misunderstand or devalue this definition completely, however, as facts are important. Facts are needed to build airplanes

and make medicines, but without the second definition, there would never be any *new* knowledge acquired.

The second application of the word science is as a verb. It can be described as all of the practices, procedures, and methods used to discover new information, test old information, or explain existing phenomena. In the past, some of these practices have often been taught in elementary school as a series of fixed steps called the "scientific method," but this is a misrepresentation of the way in which most scientists actually go about making discoveries. The Science and Engineering Practices of the Next Generation Science Standards (NGSS) more accurately represent the skills that are valued in scientific inquiry. These include asking questions, developing and using models, planning and carrying out investigations, analyzing and interpreting data, using mathematical thinking, constructing explanations, communicating information, and engaging in an argument with evidence. While they might seem similar to the linear scientific method that used to be taught in early grades, these practices do not necessarily need to happen in the order stated above, nor do they all have to happen in the same investigation. These practices form the foundation for discovering and testing information and using that information to guide us to explanations about the world. By using these practices as a tool for learning the content, students can get an understanding of not only what is being learned, but also how scientists go about learning it.

The final definition of science can be used as an adverb to describe how science, as a verb, is conducted. It describes the systematic processes involved in quality assurance of the information gathered about the universe and the attitudes of those carrying out and critiquing these processes. This is the least understood aspect of science, but the most important to communicate if students and the public are to accept what they are being taught. These characteristics and procedures are truly what separate science from any other way of knowing and allow people to have confidence in the conclusions reached. They include characteristics such as objectivity, open-mindedness, and skepticism. These qualities allow scientists to consider all relevant information while remaining wary. It also allows them to disregard any explanations that do not fit the data, regardless of personal beliefs.

The use of controlled studies and statistical analysis allow experiments to be designed in such a way that bias and most confounding variables are eliminated so that data and conclusions can be determined significant and valid. The peer review process helps to ensure that mistakes and poorly designed studies are weeded out before this misinformation is published, holding researchers accountable for quality. Meta-analyses and scientific consensus allow big ideas to be thoroughly examined, critiqued, and confirmed. For example, we do not trust that vaccines are safe and effective because of guesswork or because "Big Pharma" pays to have dissenters silenced. We trust their safety and effectiveness because they are rigorously developed and tested in multiple stages, usually taking more than a decade, by both private and government organizations. Every step of the process is critically monitored and regulated by government oversight. Even after licensing is obtained, multiple government agencies including the Centers for Disease Control and the US Food and Drug Administration, in addition to private groups, continually monitor the progress for potential problems or concerns.

Confidence is key. A scientist does not make a claim without first being confident in the significance of the data and the conclusion reached by the interpretation of the data. If the claim is more extraordinary, then the amount of evidence and statistical confidence that we have in that evidence must be higher to match for it to be accepted. This is where education has been lacking. Most K-12 science programs do a decent job of teaching the first two aspects of the nature of science. Students leave high school knowing the basics of those topics we deem important, and a decent amount also leave with knowledge of how that information is discovered and tested. Very few, however, are ever taught why they should actually have confidence in this knowledge or why they should trust new scientific discoveries and explanations. Most students do not learn about the peer review process that acts as a checks-and-balances system to ensure high-quality studies. They do not learn about meta-analyses in which thousands of studies from all around the world with billions of data points can be compiled and analyzed to come up with a single extremely confident conclusion, as is the case with genetically modified organisms (GMO) safety.

They do not learn how to filter *all* information through the lens of unbiased, open-minded skepticism. If they had learned this in school, they would know that they could trust and accept the content presented to them. They would not be taken in by things such as fad diets, healing crystals, and homeopathy. They would be able to make good decisions based on confidence that science-based practices have been tested and evaluated, and that non-scientific alternatives are at best a shot in the dark, at worst a harmful scam.

More research needs to be done about how to best implement this type of instruction into K-12 education. Should there be a formal nature of science course, perhaps in middle school or early high school? Would it be more effective to begin every science course with a nature of science unit? Or, perhaps the best way to teach the nature of science is simply to integrate it into each lesson that is being taught in all science courses. Regardless of which method of implementation works best, none of them will be successful if science teachers are not properly educated. Most teacher-training programs do not have such a course as a requirement; mine did not. This means that those of us who can share the knowledge and skills with our peers have an obligation to do so. If teachers can be comfortable and confident in not only the content that they are teaching, but also in communicating to their students how that content was discovered and why we are sure it is accurate, most problems with non-acceptance or denial of these topics would resolve themselves. Teachers can pile a mountain of physical evidence in front of a student, but if the student does not know how to analyze that evidence or why they should have confidence in it, then that mountain may as well be made of hopes and dreams for all the good it will do in convincing them.

Through years of trial and error and armed with degrees in education, biology, and library science, I have stumbled upon the teaching method that many teachers, wiser and more effective than myself, have been so successful at implementing: storytelling. I know of no better way to ignite the passion and imagination of children than with an intriguing story; and I know of no other story more intriguing than that of the history of all life on Earth. As the writer and cultural anthropologist Mary Catherine Bateson once said, "The human species thinks in metaphors and learns through stories."

Evolution and the nature of science fit together like perfect puzzle pieces. It is nearly impossible to accept the concept of evolution without a solid understanding of the nature of science; and there are no other stories that I know of in the history of science that demonstrate the true nature of science better than that of the discoveries explaining the mechanisms of evolution (Narguizian, 2004). These are stories of mistakes, adventures, struggles, personal conflict, and revelations. These are stories of debunking flawed hypotheses, shaking the foundations of religious fundamentalism, and fighting political battles leading up to modern-day court cases.

It is one thing to read about content in a textbook or listen to a lecture about abstract information. It is quite a different thing entirely to hear a story about how a discovery was made or to read a scientist's thoughts in their own words (McKinney and Michalovic 2004). The first time I read *On the Origin of Species* by Charles Darwin, I was surprised by how clearly he articulated his ideas and how humble and full of self-doubt he was. It made me feel that, if even Darwin lacked confidence and still accomplished so much, then maybe one day I could make my own contributions. When I learned about Alfred Russel Wallace's role in the development of evolutionary theory, I began exploring his unique history. Wallace's life story reads like a Robert Louis Stevenson novel full of death-defying shipwrecks and recoveries from tropical illnesses. Who needs a fictional pirate when you have a real-life adventurer who experienced it all and still managed to be one of the most impressive examples of humility and integrity anyone has ever heard of? Later, when I became a teacher, I did not think twice about retelling these amazing stories of circumnavigation and jungle exploration to my students to catch the attention of those interested in action, all of which can be accessed for free online (van Wyhe 2002, 2012).

But wait! There's more. Not interested in action and adventure? How about drama? The internal conflict that Darwin faced within himself and his struggle to find reconciliation between his scientific discoveries and his religious ideas are riveting. The openness of the relationship he shared with his wife (and first cousin), Emma, and the void she feared because of these conflicts is second only to a Jane Austen novel. Drama not for you? Try the unflinching loyalty and support that Thomas Huxley showed for Charles Darwin throughout their decades-long friendship.

On the other hand, many students are thrilled by the backstabbing betrayal of Rosalind Franklin by her colleagues in unraveling of the structure of DNA, a key piece of information that aided in the modern synthesis that bridges the gap between evolution and genetics (James Watson, Francis Crick, Maurice Wilkins, and Rosalind Franklin, 2018). Just as exciting is the scandal that Lynn Margulis caused in her fierce defense of the endosymbiotic theory and its place as a creator of new species in opposition to neo-Darwinism. Some scientists rallied behind this woman-warrior while others ridiculed her, but in the end, endosymbiosis became another piece of the extended evolutionary synthesis that seeks to reconcile natural selection, endosymbiosis, lateral gene transfer, and epigenetics into a cohesive, unified explanation for the history of life on this planet (O'Malley, 2015). Finally, who doesn't love the tale of Peter and Rosemary Grant and their decades-long study of Galápagos finches? Theirs is a story of victory and vindication for Darwin because it showed that natural selection and the origin of new species could be seen within a single lifetime, something Darwin originally thought impossible (Achenbach, 2014). In the history of the development of evolutionary theory, there is a story to suit everyone's interests.

Do not pass up opportunities to share these stories with students. Giving faces and voices to the scientists who meticulously gathered evidence for evolutionary theory allows students to connect with these historical giants and experience their journeys of discovery vicariously. Reading excerpts from their journals and their personal letters to friends and colleagues helps students to get into their minds and empathize with their internal conflicts. These primary sources and biographical viewpoints show students that even the greatest ideas in history are attainable by people who experience confusion, doubts, tragedies, and insecurities just like them, which can give students the confidence to make mistakes and keep on trying. By sifting through real historical data, students can use the techniques of the scientists themselves, draw their own conclusions, and compare them to the current

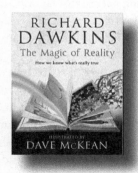
scientific consensus. Learning how evolutionary theory has been challenged and critiqued from every possible angle throughout the last century and a half, all while growing stronger and becoming more refined, can show students why the modern scientific community is so confident in its place as the foundation for building the fields of biology, agriculture, medicine, conservation, and even psychology.

As that foundation, these two ideas of evolution and the nature of science should be introduced early in curricula and used as the binocular lenses through which all other concepts are viewed. By using stories of the discovery and development of evolutionary theory to teach the nature of science, we could see a class full of students eager to learn more instead of a class full of disengaged doubters and deniers. With this method of science education, we could see a generation of skeptical consumers of information and thoughtful voters who use evidence to drive policy decisions. We could see adults who consult a doctor instead of a social media horoscope when they are sick. We could see a society, having learned empathy from these stories, that is tolerant of diversity, compassionate toward those in need, and accepting of new thoughts and ideas all based on a true understanding of the nature of science, evolutionary theory, and scientific literacy.

Throughout my career in the classroom, my passions for science and literature have only grown, and recently I have found joy and satisfaction in sharing these passions with other teachers as well as my students. I firmly believe that teachers are one another's best resources. We all walk into slightly different classrooms with slightly different students who have slightly different histories, but at the core, students are students. They all need love and encouragement. They are all curious about themselves and the world around them. They are all fascinated by stories. Being a teacher leader and ambassador for organizations like Teacher Institute for Evolutionary Science (TIES) and National Center for Science Education (NCSE) has given me the opportunity to share my passions, skills, and stories with amazing teachers from all over the country. Presenting lessons, strategies, and resources at workshops and conferences has been both a rewarding and humbling experience. I have had the chance to have conversations with teachers from everywhere in the United States about topics ranging from science and evolution to classroom management and building meaningful student/teacher relationships. I have shared ideas with first-year science teachers, retired science teachers, librarians, administrators, and superintendents.

Everyone I meet wants the same thing for their students: success. Organizations such as TIES and NCSE provide that opportunity. They give aid to teachers who want their students to be successful, but may not have the knowledge or skills to make that happen without assistance. These organizations encourage teachers to be leaders in their professional learning communities and provide them with resources to collaborate and share with their fellow teachers. We all have strengths to share, and we all have room for growing and learning. I have been fortunate to have had amazing mentors in my career who have guided me and given advice, and I look forward to expanding my growing network of fantastic colleagues to both learn new ideas for my own classroom and give back to this wonderful teaching community in any way possible.

Introducing Evolution

Almost all of my students (and adults that I speak with) have misconceptions about evolution. These myths, misconceptions, and sometimes outright lies about the process of evolution are perpetuated by the media, newspapers, movies, religious leaders, and magazines. While I know that some of these issues are created or passed along on purpose to divide and polarize the public, the vast majority of these issues are not done maliciously and can be boiled down to ignorance, innocent mistakes, or attempts to create sensationalized headlines in order to grab people's attention. Science magazines can sometimes be the worst perpetrators of these "high crimes" in their attempt to make every article that they publish seem like a groundbreaking, paradigm-shifting discovery in order to sell subscriptions.

In today's modern world in which children and adults alike are overexposed to #ClickBait, it is more important than ever to demonstrate to our students how easy it is to fall prey to false or misleading information. I take a two-pronged approach to accomplishing this task in my classroom. First of all, I always begin each school year with a nature of science unit in which we discuss the various aspects of what Science is, how we acquire knowledge, and how we validate knowledge to ensure its reliability (more on that in a later section of this chapter). As part of this nature of science unit, one of the things that I teach my students about is the CRAAP test ("Is this source or information good?" 2019). My students and I love the CRAAP test for many reasons; the least of which is that we all get to say "crap" without getting into trouble. More importantly, however, the CRAAP test is a way to ensure the reliability of sources, especially the online sources that they will most often encounter in their daily lives.

The acronym CRAAP stands for Currency, Relevance, Authority, Accuracy, and Purpose. The test comes with a checklist of questions that students should ask themselves about each of these topics when they are evaluating any source of information for validity or reliability. The students find this useful both for my class and outside of my class when they are browsing social media, watching the news, or critically analyzing marketing advertisements for products. The CRAAP test teaches them to consider things such as what is the source of this information and can it be trusted, who wrote this and why, how old is this information and has it been updated, can I check this information against another source to make sure it is accurate, and are they trying to persuade me of something, sell me something, or just inform me.

After introducing the CRAAP test along with the nature of science unit, we constantly practice using it with any source materials that we come into contact with until the students no longer need the checklist and can just perform the necessary evaluations on their own. This helps to transform my students into skeptical, critical consumers of information in all aspects of their school and home lives. This practice also helps when we get to our dedicated evolution unit later in the year.

One of the activities that we do in our evolution unit builds off of this critical skepticism with a game called Spot the Misconception (like the example on p.92). During this activity, I show my students headlines involving evolution-related topics from various news sources, science magazines, and websites. My students are tasked with working in groups to figure out what the misconception is, why it is a misconception, and then rewriting the headline so that it no longer contains a misconception or is no longer misleading.

This activity is useful because it helps to accomplish many different goals at once. It gets them to think about, discuss, and correct misconceptions that they may have heard or held themselves about evolution in a non-confrontational way. It helps them to evaluate why misconceptions exist and how they get spread from source to source or person to person. It helps them reapply what they have learned about both the nature of science and evolution by asking whether the misconception is about how Evolution itself works or just in a misunderstanding of the process of scientific

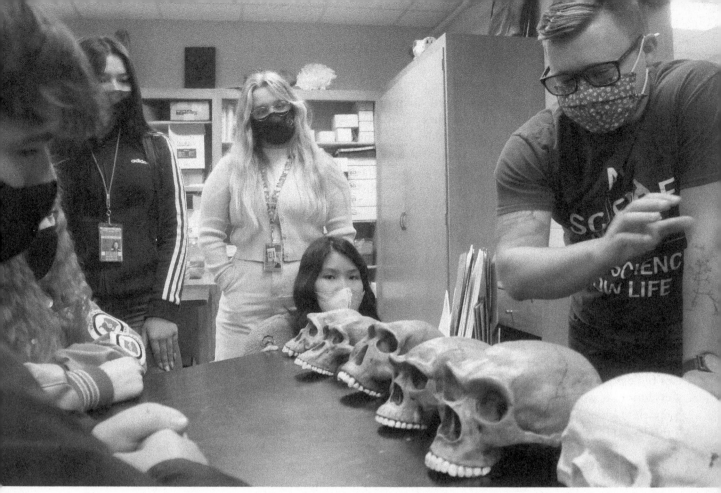

Blake Touchet and students discussing hominin evolution before a skull morphology lab.

knowledge-making and validation. Finally, it reinforces the idea that they must remain constantly vigilant against misinformation even when it comes from a source that most people would consider reputable (I'm looking at you, *New Scientist* and *National Geographic*).

Favorite Investigative Unit or Activity

The best activities for engaging students in evolution are those that are relevant to their personal lives, demonstrate the basic principles of natural selection, show the explanatory power of the theory, and use evolution as a unifying concept for all aspects of biology. In my personal experience, the activity that checks all of these boxes best is "A Rainbow of Sepia" (Prud'homme-Genereux, 2018). This activity is a fully NGSS-aligned case study published in the National Center for Case

Study Teaching in Science database that allows students to investigate the question "Why are people different colors?" through the case study stories, maps, charts, and graphs as well as videos and resources from HHMI's "The Biology of Skin Color" BioInteractive unit (2016).

Box number 1 (relevant to students' personal lives): Check! There isn't a child in the world who has not asked the question "Why are people different colors?" Every student is interested in the evolution of skin color whether out of pure curiosity or because they have seen the effects of oppression and privilege that can be culturally assigned to people with certain skin colors. Students want to understand the reasons why they are different from others in such a visibly obvious way. Additionally, if this case study is taught correctly it can open a dialogue into not just scientific questions, but also questions of social justice, empathy, and what it means to be a good human in our diverse modern society.

As students work through the activity, they are also working through the scientific process of questioning what the selective pressures could possibly be that have driven the evolution of different skin colors across the globe. They look at maps showing UV intensity and skin color gradients, analyze graphs showing folate and vitamin D levels in the blood, and ask questions about disease, nutrition, and cancer. Throughout the case, they are acting as investigators and enjoy a sense of discovery as they learn about each new layer that influences fitness in populations at different latitudes and elevations.

Finally, I love this case study because it ties in the anatomy of the integumentary, skeletal, and reproductive systems with gene expression, ecology, and modern medicine all through the lens of evolution by means of natural selection! To me, this is the perfect demonstration of how evolution anchors and connects all aspects of biology. If I had to choose only one activity to teach evolution, it would absolutely be this one. I cannot see myself teaching another year in the future without using this lesson in some form.

Dealing with Conflicts

As science teachers, we happen upon a lot of misconceptions . . . like . . . a *lot* . . . *so many misconceptions!* I encounter misconceptions almost every single day from students, and if there is ever a day when a student doesn't bring one up in class, it's almost a guarantee that I will hear one on the radio on the drive home or see one on social media before dinner that has been shared by a friend or family member. So, how do we counter this misinformation in the face of all of these overwhelming odds?

To begin, there are two major techniques that I use to ensure that I don't fall prey to misconceptions, myself, or unknowingly spread them to my students or my friends. First, I try to remain constantly critical of the information that I consume. I use techniques such as the CRAAP test to evaluate sources for validity, reliability, and bias. I question myself and my prior knowledge to see if it fits with the new information that I am reading about or learning from others. If it doesn't fit, then I dig deeper to see if it is the new information that is inaccurate or my prior knowledge that was incomplete or somehow wrong. Secondly, I attempt to spot logical fallacies in arguments or presentations and avoid falling

into these fallacies myself. To know when someone is cherry-picking data, using a strawman argument, setting unreasonable expectations, or throwing out a red herring is a powerful weapon in the fight for stopping the spread of misinformation and misconceptions. Armed with this critical skepticism, I teach my students to use all of these tools on a daily basis as well. This arsenal comes in handy when approaching misconceptions that students hold that could easily turn into a huge classroom controversy, but these tools cannot be wielded without precise skills and techniques.

When approaching misconceptions that students have, the first step is to put them at ease. I usually do this early in my class. Students need to feel that my classroom is a "brave space." Not a safe space, mind you, but a brave space. The phrase "safe space" brings to mind a place when any student can say anything without any worry of being harmed, ridiculed, laughed at, looked down on, or otherwise made to feel uncomfortable in any way (Arao and Clemens 2013). I'm sorry to say, but that ideal safe space does not exist in any classroom or school anywhere in the world, and you cannot lie to a student and tell them that they will never experience any of these things. They *will* experience them, as much as we try to minimize them, and when they do figure out that you have lied to them about your classroom being a safe space, they will shut down and you will lose them.

The alternative is to create a brave space. A brave space is a place where students are taught to share their opinions, to respectfully disagree with one another, to argue and debate in a scientific manner using evidence to support positions. A brave space is a place where students know they will be challenged to work and to think, and of course, sometimes thinking is uncomfortable, but that's OK! They may be required to learn about things or consider ideas that don't quite fit with their world-view, and that is going to cause discomfort, but that's OK! They may make a claim that turns out to be wrong or ask a question that's silly, but guess what, *that's OK!* This is what it means to be an active participant in a scientific community, which is what I want from all of my students.

One way that I put my students at ease at the beginning of the school year when I'm establishing my brave space is to show my students that everyone holds some misconceptions, even me. Part of the process of science that we have to be comfortable with is throwing out old ideas that do not hold up to deep scrutiny or new data. I usually demonstrate this by telling a personal anecdote. When I was very young, probably about four years old because I hadn't yet started school, I vividly remember a time in which I had a personal misconception exposed and then clarified.

I grew up on a little farm and would wake up early with my dad and grandfather every morning to check on our animals. We would make our rounds from the horses to the cows, sheep, pigs, chickens, and rabbits filling feed troughs and topping off water containers. One morning, I asked my dad something that had apparently been bothering me for a few days. "Dad?" I asked. "Why does it rain every night but not every day?" My dad looked at me with a very confused face and asked, "What?" I repeated the question, and after getting the same look from him and a response of "It doesn't rain every night," I added my logic. "It must rain every night because the grass is always wet in the morning when we get up to check the animals, but it doesn't rain every day." This drew a laugh from him because he finally understood what I was asking.

My misconception was a simple one. I assumed that because the grass was wet, it must have rained, and because the grass was wet every morning, it must, therefore, rain every night. I had attempted to explain a phenomenon in a logical way with the knowledge that I possessed. What I didn't know at the time was that my logic was solid, but my knowledge was incomplete; thus resulting in my misconception. After my dad explained that dew was caused by condensation and gave me the example of a cold can of soda "sweating" when you left it out of the refrigerator, my knowledge gap was filled in and my misconception was overturned with better, more complete knowledge.

My students always love this story. They get to laugh at my naivete, and I get to break the tension leading into a set of interactions that could otherwise be seen as a power struggle between them and a "know-it-all" teacher who is trying to destroy their belief systems and world-views. Instead we get to have a scientific discussion about what misconceptions they may have about

scientific topics, why they have those misconceptions, and what we can do to address them and replace the myth with reality.

Once this relationship has been established, any misconception that comes up in class can easily be addressed through a series of steps. First, I always treat each misconception with sincerity. If a student asks you if the earth is flat or spherical, you may be tempted to laugh. Don't. Even if the kid is joking or trying to get a rise out of me, I can still use it as a teachable moment to demonstrate the process and power of science. And if they asked the question, it's probably a safe bet that at least one student in the class actually does have that misconception. Second, I always try to find out what the root of the misconception was. Did they hear it from a movie, another student, their parent, their pastor, a politician, another teacher, or was it just a "dew" sort of confusion? This will help inform how delicate I am when clearing up the misconception. Nobody likes to be told that they are wrong, but everyone *hates* to be told that their mama is wrong. They will fight if their mama gets disrespected. Also, if I find out that it was a teacher that was spreading a misconception, that means that I have to have another conversation after class, and it is even harder to talk to most adults about misconceptions because they are much more stubborn than children. Finally, I always try to walk the student through both the misconception and the correct information and logic behind the concept. If they can see where their logic was flawed or what information they were missing and compare that to the full, correct concept, not only will they gain comprehension, but they will be less likely to fall back into the same type of misconception trap in the future.

As the years have progressed, I have probably heard almost every misconception that exists about biology. That makes it easier for me to predict which ones are likely to come up and what I need to do to clear it up. I like to make it a point to discuss common misconceptions with other teachers during professional developments. It helps warn them about what they might encounter and give them some ammunition against them. It also gives me more knowledge about misconceptions I have not encountered before. Every once in a while, I'll have an off-the-wall thinker who will surprise me with a new one, but that just keeps me on my toes.

Let's Play . . . Spot the Misconception! —Examples!

Resources:

https://evolution.berkeley.edu/evolibrary/misconceptions_faq.php

http://pages.nbb.cornell.edu/neurobio/websterlab/10%20Misunderstandings%20-%20SHORT.pdf

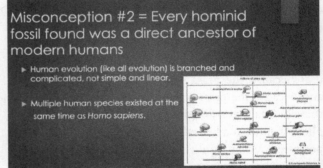

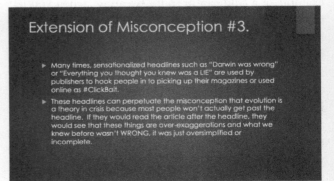

12 Evolution and a More Just Society

David Upegui, PhD
Central Falls, RI

As a child, I remember the confusion I felt about my world. Mother's dark complexion, her heavily pigmented hair, her black irises, her short stature, her gender, her working class, her thick accent, her migratory status—these all pointed to a woman with little value to the American society. She did not reflect the image the media portrayed—she was the "unwanted other." But to me, my mother was the world. I saw her strength, her tenacity, her determination—I saw a woman who was willing to work all day in a textile factory to nurture the American dream in me. Of course, I was confused; my mother did not represent society's desirable traits, yet she was the most inspiring person I knew. My mother's merits were dismissed by society because of unchangeable traits. All these conflicting ideas came crashing as I prepared to do something no else in my family had done, attend college. It was there during my first biology class that I found a new way to look at the world: evolutionary theory. With evolution my mother's characteristics made sense. I could use science to discern between natural variation and unjust social structures made by humans. Evolutionary theory not only provided a cohesive way of organizing the biological world, it also freed me from the anxiety that I was not enough. I began to see my mother's characteristics, including gender, not as a "weaknesses," but an integral and beautiful part of a species that found the courage to migrate within and out of Africa and spread throughout the planet. There was a clear contrast between the homogeneous idea of beauty and fitness as presented by the mass media/marketing complex and the population fitness provided by diversity. One is a simplistic idea designed to sell things, while the other is true fitness in the evolutionary sense. One is shallow and has been used to dehumanize, while the evolutionary fitness seen in diversity aligns much better with biology and real-world experiences. As such, with the realization of evolution, students can appreciate and reveal a broader set of truths that can open up a rigorous questioning mentality that is transformative for them as they figure out who they are and their "place" in society.

To say that science became the exit ticket out of poverty is an understatement. Science became a way of understanding our universe and in the process to view every aspect of life differently—thinking critically. My professors and mentors turned my capricious curiosity into a systematic and focused methodology. The scientific skills I honed in school undoubtedly prepared me for my graduate work and later my employment at an Ivy League university. I was fully appreciative of the benefits of academic advances and the rewards of intellectual development. Science education, and in particular, evolutionary theory had given me a better grasp of the world and my place in it. Then my first son was born—a perfect little boy, with Down syndrome. In genetics courses, I had certainly read about trisomy 21, but I never suspected that this six-pound wonder would change my world so quickly. It was his birth that began changing my heart about my own future and turned my ideas of being an epidemiologist into something new. After some struggle with these new perspectives, I decided to earn a high school teaching certificate where I could channel my passion and knowledge of science into a rewarding experience for others. Even after it was complete, I was hesitant to leave my job and take a significant pay cut in order to try something where I would be a rookie. Then the paradigm shift came along with unimaginable events, my high school alma mater (the poorest school in Rhode Island) was in disarray—all teachers were fired and the conflagration was constant. Once I walked into room 132 where I once sat as a science student, all hesitation vanished. I was home, and this time I was armed with knowledge. I would begin to teach others about evolution and how it can help us to gain perspectives about our own lives. Since that day, I have been translating the content of the textbooks into lessons that engage and excite my students. With limited resources but an overzealous attitude, I began to incorporate all my background and expertise as well as inviting experts into the classes. As it turns out learning about evolutionary theory can be the ticket out of more than monetary poverty, it can help us to create a more just society.

Challenges of Teaching Evolution in the US

I see science as an expansive web of knowledge that eventually finds connections between experiences; and as a method for exploring and appreciating our universe, science requires us to recognize and create these connections. Therefore, it is imperative that students build on their prior experiences and knowledge. This is a profound task that can be difficult given the broad range of backgrounds that students bring into the classroom, but one that must be constantly pursued if the content is to become real for the students. In my experience, when students are regularly encouraged to participate and reflect upon their own learning, they become agents of their own academic development.

E. O. Wilson (2012) wrote: "Humanity is strengthened by a broad portfolio of genes that can generate new talents, additional resistance to diseases, and perhaps even new ways of seeing reality. For scientific as well as for moral reasons, we should learn to promote human biological diversity for its own sake instead of using it to justify prejudice and conflict." As a biological

statement, he might have added, as well: "for our species' sake." Biology education that emphasizes how genes interact with the environment by favoring certain traits which help species perpetuate clearly demonstrates to students that their physical differences are natural and may result from many things, including genetic responses to environmental pressures. For example, by understanding that the current biota—including themselves—is but an arbitrary slice of biotic history, students can see color pigmentation not as a way of separating humans in socially discriminatory ways, but instead as the result of differences based on the amount of ultraviolet radiation to which their ancestors were exposed on a regular basis (Jablonski, 2010). In this way, darker pigmentation is a historical inheritance, rather than a class judgment. Therefore, by using science (namely, evolutionary theory), we can appreciate that race as a subject of study presents an invaluable opportunity for students to explore how science actually works and how it has been misused in the past.

On the other hand, culture is malleable and changeable via education. Hence, in order to get to a more equitable society, we should leverage that education to promote a culture centered on social justice. Henrich (2016) provides evidence to suggest that language complexity, as demonstrated by the number of vocabulary words, is closely related to cultural complexity. If we provide learning opportunities for children to utilize their language constructively, we afford children avenues to enhance their total development, and through their consequential behavior, our society. It may surprise us how critical thinking skills will benefit us and our descendants, the future stewards of this world. Ironically, in the early 1900s, the evils of social Darwinism were attributed to evolutionary theory; my experience is that when all students are given the intellectual tools to analyze and to question meaningfully—tools that we can begin to give them through the introduction of evolutionary theory—they are also being given the tools to create a just society.

Within biology, evolutionary theory offers students opportunities to understand the connections between all biological life on Earth and phenotypical variations as the direct result of natural selection. In this light, concepts such as "race" can be challenged. The use of the concept of race focuses on physical differences as significant and

Central Falls High School AP biology students completing a phylogenetics lab activity using mammalian skulls with David Fastovksy (URI) and David Upegui.

in a racist society, these ideas are then used to validate unjust social structures (Vance 1987). For example, British educational psychologist Cyril L. Burt argued that intelligence is inherited and not the result of the environment (Burt 1909). As it turns out, in making his case, Burt fabricated statistical evidence which falsely demonstrated that European races were superior to "savage races." This type of overt racist utilization of science exemplifies the power of science and the ability of those with the power to abuse and oppress those from marginalized communities.

Vance (1987) promotes the role biology can play in establishing a framework for challenging racist images and practices. These challenges, in turn, become politically powerful statements which can ultimately better society. To this end, Vance lists stereotypes assigned to Black people (i.e., Black people have stronger body odors, are more aggressive, they look more alike, and are more closely related to apes) and then describes how anti-racist science education can challenge these beliefs to rectify these misconceptions. In other words, without a deep appreciation for scientific methodology and knowledge about natural selection and evolution, the aforementioned statements can be erroneously seemed to be validated by science. Therefore, a true understanding of biology, evolution, and scientific methodology can become a tool for challenging ignorance and increase political efficacy. Angela Calabrese-Barton (2003) expresses this idea directly: "To know science—and to be deemed as one who knows science—is a uniquely powerful stance. Science education is political. It promotes particular images of power, knowledge, and values by rewarding particular forms of individual and institutional behavior."

Evolutionary theory affords students the ability to focus their thinking by allowing them to integrate the historical dimension of the modern biota, rather than just memorizing disjointed parts of the domain, whose ultimate unity is to be found in their evolutionary history (e.g., Dobzhanski, 1973). Biology notwithstanding, it is pivotal that our students have every possible opportunity to think critically. Given the fact that evolutionary theory is still controversial, its introduction provides a great opportunity to explore scientific philosophy, methods, and its ultimate social impacts.

We are standing at the best time of history for the human race. Our technology and creativity have allowed our large brains to gain and retain tremendous amounts of knowledge. We have accomplished so much because we are, as Newton said, "standing on the shoulder of giants." One of the most prolific gifts we have been given are the ideas and theories of evolution. To understand our lineage and connection with the greater cosmos has been a pursuit of all civilizations, and today we are separating fact from fiction thanks to the utilization of scientific processes. That's what is fundamentally beautiful about current evolutionary theory—it stands the constant flow of data and it best explains how Earth has come to be this "pale blue dot" full of life.

To teach biology (the science of life) is to teach evolution; without it nothing else makes sense. Every concept from biochemistry to plant/animal physiology is intertwined with the pressures and effects of evolution. Most of my students come to me with limited ideas of evolution and many misconceptions. I feel it is my job to show them how evolution is the most reasonable explanation and most sensible connection to life. Evolution permeates every aspect of my teaching and it even finds ways to educate others about seemly mundane dilemmas. I teach in the school with the highest percentage of students living in poverty in the state of Rhode Island, and because of this, I feel that teaching evolutionary theory empowers my students in ways that other subjects cannot. For example, when one of my students was confronted by another teacher about "migration" to the US, I had her reflect on the natural migratory patterns of other living organisms and how that relates to fitness, adaptations, survival, and finally passing of genes. After she released her anger, she realized that this man's

ignorance was not just about US borders/laws but more fundamentally about biology (life)!

When I left my research position to teach at my high school alma mater, I knew that I would be facing formidable challenges. Some of these included a graduation rate of less than 50 percent, low teacher morale, and dismal standardized test results. All of these were further compounded by the ugliness of poverty in this urban district. However, none of these were as powerful as my colleagues' dedication, and intentions. The following year, I was fortunate enough to begin teaching AP biology. Some critics believed that I was setting myself and my students up for failure. However, I believed in the students and furthermore I decided to "call in the cavalry." Every year, we have had visits from experts and professionals that include Ken Miller from Brown University, who spoke about the Dover trials; Lloyd Matsumoto from Rhode Island College, who presented on enzyme activity and evolutionary pathways; Brett Pellock and Cara Pina from Providence College, who brought microbiology and evolutionary expertise; and David Fastovsky from the University of Rhode Island who led discussions on evolutionary adaptations and phylogenies.

In addition to the typical phylogenetic tree creation and Hardy-Weinberg laboratories, I have focused on connecting evolution with every topic throughout the year. For example, when I introduced the concepts of genotypes and phenotypes, I brought my family dog for a visit. The students loved interacting with the animal, but most importantly each of them spent several minutes trying to figure out his genetic background by looking at his physical characteristics as well as his behavior (his mother was an Australian shepherd, and he demonstrates herding behaviors). Also, free aquarium visits for our class allowed me to emphasize physical adaptations that different fish have developed over millennia. Each student "created" a fish using the major anatomical structures (mouth type, body type, fins, coloration, etc.) and during our visit found a similar fish which they described in detail including biological niche and connecting form and function. During our class, I emphasize the endosymbiotic theory and I ask students to support it by using all current data. Most importantly, I ask them to think in terms of evolution and the importance that these events have had on life on our planet. It

BOOK RECOMMENDATIONS

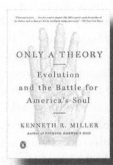

There is a voluminous collection of wonderful books about evolutionary theory. I strongly suggest reading a general book about teaching evolution such as *Only a Theory* (2009) by Ken Miller. Moreover, I have several suggestions that will complement our general understanding about evolution and social justice. The first of these books is R.C. Lewontin's *Biology as Ideology* (1991) and although this book predates the Human Genome Project, much of it still relevant and significant. For instance, he writes: "Science is more than an institution devoted to the manipulation of the physical world. It also has a function in the formation of consciousness about the political and social world. Science in that sense is part of the general process of education, and the assertion of scientists are the basis for a great deal of the enterprise of forming consciousness. Education in general, and scientific education in particular, is meant not only to make us confident to manipulate the world but also to form our social attitudes."

The next book that I want to recommend is Dr. Ann Morning's *The Nature of Race: How Scientists Think and Teach about Human Difference.* (2011). In this book, she describes the history of race science in the US. She writes: "Our concepts of race are not limited to abstract definitions but rather incorporate a wide range of notions of what race is, what distinguishes one race from another, how many and which races there are, how we can discern an individual's race, and how or why races emerge. In short, racial concepts are working models of what race is, how it operates, and why it matters . . . The dispute has frequently been translated as a question of whether race is 'real' with the understanding that biology represents reality and social facts denote fiction."

Agustin Fuentes, in *Why We Believe: Evolution and the Human Way of Being* (2019), outlines the power of evolutionary theory and differentiates between the social construct of race and biological-related groupings of humans.

Graves Jr. and Goodman's *Racism, Not Race: Answers to Frequently Asked Questions* (2021) describes how evolutionary theory shows us that human genetic variation is real but it is absolutely not the same as the commonly held idea of race.

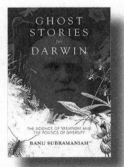

Finally, the last book I recommend is Dr. Banu Subramaniam's *Ghost Stories for Darwin* (2014). In her book Subramaniam dives deep into evolutionary biology and social structure. She writes: "Contrary to the stereotype of the 'sterile' scientist producing 'pure' knowledge about the natural world, the history of evolutionary biology reveals a rich history of scientists concerned with 'social problems' as they understood them."

is tremendously satisfying to recognize that students are making these connections and demonstrate it by creating powerful artifacts (i.e., presentations, videos).

For my students, education is more than just "what happens in school." It is the key that unlocks better futures and enables them to put their dreams into action. I see biology as the most complete science and evolution as the axis that centers all other ideas. If our school is a microcosm of the world, things are looking up and from my point of view, science education is the force making the difference. Evolutionary theory is the way to connect all of biology and also prepare my students to think critically and examine their own ideas.

I have done some things which have paid huge dividends in my students' learning. None of these are novel or original, but in conjunction, they have worked for the particular student population I teach. Here are some general practices that have afforded my students with powerful and meaningful opportunities:

1. **Open the classroom.** By this, I literally and metaphorically mean extend the walls of the biology classroom. Bring people into the classroom so that the students understand that biology is a "universal endeavor" and the kinds of habits of mind which we practice are applicable to many other circumstances. Our classrooms may be small, but the ideas we handle are monumental. Students need to see that. It is also important to keep up with current biological ideas and theories, and I have a dedicated time of open discussion with my students. They are eager to teach me as well.

2. **Call in the experts**. Regardless of our own background, shortcomings can be best addressed by experts. Over the years I have made SOS calls to everyone from the author of the most popular high school biology textbook (Ken Miller), to my own college professors (Duncan White and Lloyd Matsumoto) and also teachers within our school. I was pleasantly surprised to find out that most people are willing to help. Whether it is with equipment (such as Spec 20s), time to meet with me, or actual classroom interactions, these collaborations have proven extremely important. I am always moved (though not surprised) that every visitor leaves feeling good about their interaction and about the students.

3. **Use resources** (especially electronic ones). Collaborations do not have to be limited to regional geography. Some of the most dynamic lesson plans I found have been shared with me by other biology teachers far from Rhode Island. Organizations such as TIES, the College Board, the National Association of Biology Teachers (NABT), and the National Science Teachers Association (NSTA) are places to share and gain insights from other teachers. However, these come with a caveat: Even great lesson plans from others *must* be adjusted and personalized for the students being served. Taking a resource and using it ad hoc can make the students (and the teacher) feel lost. I have also found some responsive educators willing to share amazing stuff such as Paul Andersen, David Knuffke, Cheryl Hollinger, Lee Ferguson, and Kirstin Milks. Moreover, they are all responsive and amazing. Lastly, the web is filled with truly empowering materials and it can be a gold mine. I have found that organizations such as NESCent, the National Center for Case Study Teaching in Science, ENSI/SENSI and HHMI are bursting with valuable and pertinent materials.

4. **Have a purpose for *everything***. Every conversation, every topic, every idea can be explored, expanded, and enriched—if there is a purpose. For example, before traveling on any field trip, the students must focus their critical thinking and observational skills so that the experience is "realized" to them. Stating the purpose often serves to guide my students and prepares them for their work.

5 Most importantly, **return the favors**. When I began teaching high school, I did not feel that I had much to contribute to the experts (educators and other professionals) who were helping. However, I now know that there is much we can do for each other. As we open ourselves to others, others open themselves more to us—symbiosis!

Recommendations

Given the current circumstances in our country in which data, facts, and science are challenged by charlatans, it is imperative that we teach science to our students. The recent sets of science standards (both AP biology and NGSS) offer many opportunities for us to rethink

VIDEO RECOMMENDATIONS

Many wonderful videos enable students to visualize evolution (including the difficult concept of time on Earth) and have already been discussed thoughout the book. Therefore, rather than list one of those, I will discuss two videos that show students the importance of learning about evolutionary theory. To begin, I show the video *Should Evolution Be Taught in Schools? Worst Top 15 Miss USA Contestant Answers*. Subsequently, I show the video which mocks the first responses *Miss USA 2011—Should Math Be Taught In Schools?* (it makes all my students laugh).

science education and, in particular, evolution. Both sets' standards place value on evolution: at least 25 percent of the NGSS Life Science standards address evolution directly, as is also the case with the AP biology curriculum. Therefore, an area that needs to be addressed is the lack of understanding, appreciation, and commitment to these standards that place a greater emphasis on students utilizing science, rather than just memorizing facts. To this end, professional development (PD) is needed for both the administrators and the teachers. This PD must include a clear delineation of the fundamental theoretical shifts in thinking, as well as the application of sample lessons. During the sample lessons, participants will compare and contrast the traditional ways of running science exercises with the newer ways encouraged by the NGSS and the College Board.

Lastly, since in theory, both standards value the students' culture, it is at this point where social justice issues can be injected into the curriculum. The infusion of relevance must come from a dialectical experience where students and teachers inventory the local environment for a sample of issues that can be explored via science. For example, local environmental problems can be explored using science (i.e., Which chemicals are causing problems? What tests do we perform to find out safe levels of chemicals? What effects do these unsafe chemical levels have on the biota?).

When lessons are directly connected to the lives of the students, I have experienced students to be engaged to a greater degree. Furthermore, teachers will benefit by

having less disciplinary issues since students will likely be focusing on solving problems. I also see direct benefits for the community, where students can become advocates and problem-solvers. The empowerment that comes from students taking action will increase their self-efficacy and result in the better applicability of academic knowledge and skills.

The second recommendation is for the development of specific lessons that include ways to explore pertinent and general social issues and unjust structures. For example, rich and exploratory science lessons can be focused on race, gender, and/or migration and the values and unearned privileges society places on certain individuals. Moreover, these lessons should expand outside the science classroom and become multidisciplinary in nature so that the students are exploring these issues in their social studies, art, and English language classes. At the end of each lesson, teachers and students will be required to reflect, in writing, on the sociocultural relevance of the curriculum.

As mentioned earlier, a direct benefit to students will be the deeper appreciation of race as a non-valid category for separating humans. Once students appreciate that skin pigmentation levels are related to where on the globe their ancestors spent their lives and that all humans migrated out of Africa with dark pigmentation, students will come to appreciate that attributes about intelligence, economic capacity, and academic performance are the result of social constructs and not inheritance. In much the same way, students can learn that many biological entities naturally migrate (from monarch butterflies to whales) to search for better environmental contingencies and that humans are a migrant species, students will come to see that rhetoric about migrants is not founded on biological/natural concepts. Lastly, by exposing and emphasizing the contribution of women and minorities to the development and progress in science (i.e., Abu Ali al-Hasan, Lynn Margulis, Rosalind Franklin), students can begin to understand that socially constructed ideas have limited the access and appreciation of non-dominant peoples in science.

Introducing Evolution

For me, evolutionary theory is the core and connector of all of biology, therefore, the introduction to the biology class is the introduction to evolution. At the beginning of the school year, I want to ensure that students become aware of how our class will co-construct knowledge. In that light, it is important for students to know that the basic required guidelines for our class to be successful. The rules are very simple: be prepared, be present, and be respectful. They apply to everyone in the room, including me (and this is stated). It is our way of agreeing on how we can be at our best with each other as we learn.

In the same vein, students need to know that much of my teaching is anchored on storytelling and inquiry. For example, the concept of human races presents an interesting and engaging phenomena. There are differences in groups of individuals, but how did these differences become the idea of races; and how did races become a way to segregate and oppress some people? Herein lies an opportunity to use a specific lesson(s) that address liberation (or empowerment). Namely, I introduce the science (or lack thereof) underlying the notion of human races. Skin coloration has historically been used to segregate and discriminate against people, but what if we take a look at the data of ultraviolet radiation (UVB) and human skin pigmentation patterns? Students begin to see that skin coloration is based on where our closest ancestors lived: the closer to the equator, the more skin pigmentation they had (as a natural protection from damaging sun rays). Once students appreciate that natural variation, they can begin to question why skin pigmentation was erroneously connected to human capacity.

My students leave my class knowing that we are in fact only one human species (otherwise we could not successfully breed and have viable offspring). Undoubtedly, there are tons of lessons in all content areas that can be developed, delivered, and shared with our students—we just need to consider them as what they truly are: the future stewards of the earth.

Favorite Investigative Unit or Activity

Given that evolutionary theory is directly discussed and explored throughout the year, no single activity, but rather several activities are incorporated into every unit. The beginning of the year is full of activities that expose the students to evolution at work. For example, the Howard Hughes Medical Institute's *Phylogenetic Tree of Anole Lizards* presents a great story that allows students to explore the phenomena of these organisms. Similarly, the University of California Museum of Paleontology's Understanding Evolution website has several useful stories, including "Survival of the Sneakiest," where sexual selection is explored. Following several of these case studies/stories in nature, students will collect their own data through an activity such as the "thumb" and/or several of the activities provided on the TIES website, such as the moth simulation and the bird lab. Lastly, MIT's *Blossoms* video on variation has an easy-to-follow set of NGGS-aligned activities: https://bit.ly/variationessential.

Once we progress through the curriculum, other opportunities arise for the exploration of evolution. For instance, as an introduction to biochemistry, we may eat (offered to students with parental consent) hot peppers and discuss the evolution of capsaicin. During microbiology, we cover antibiotic resistance and viral mutations. With

Yearlong Biology Sequence

1. Introduction: foundations of biology
EVO PRACTICE: What is biology? Why study it? How does "every" living thing connect?

2. Evolutionary theory and natural selection
EVO PRACTICE: Who is Charles Darwin? What was happening historically that enabled him to come up with this publication? Why do we support his theory of evolution through natural selection?

3. Tour of the cell (membrane transport/signaling)
EVO PRACTICE: How did cells evolve? What does the timeline for cells show about complexity? What is endosymbiosis, and how does it connect to evolution of complex organisms?

4. Metabolism (cellular respiration /photosynthesis)
EVO PRACTICE: Why is energy so important for evolution? How does evolution align with thermodynamics?

5. Cell cycle
EVO PRACTICE: What preceded the cell cycle, as we know it, in evolutionary terms? How do we know?

6. Meiosis and sexual life cycles
EVO PRACTICE: What does sexual reproduction afford/cost organism in evolutionary terms?

7. DNA/RNA/Protein gene expression
EVO PRACTICE: What pressures may have influenced the evolution of molecular systems? What was the evolutionary sequence of information molecules?

8. Genetics Mendel and gene idea
EVO PRACTICE: How has evolutionary theory changed since the exploration and data from Mendel?

9. Population ecology and the distribution of organisms (food webs)
EVO PRACTICE: How do interactions between species affect evolutionary pathways? How does variation within a population help species survive?

10. Diversity of Life (prokaryotes/eukaryotes), plants, fungi, animal diversity
EVO PRACTICE: How does the history of life on earth reflect complexity of organisms?

11. Physiology homeostasis and endocrine and nervous and sensory systems
EVO PRACTICE: How does multicellularity afford organisms greater homeostatic control?

What evolutionary pressures shaped the development of humans and our large brains?

cellular biology, we discuss Lynn Margulis and the endo-symbiotic theory. For the ecology and human impact units, there are plenty of examples of how evolution plays a significant role. It is important that at the conclusion of each unit students are provided opportunities to explore the connections with evolutionary theory and explicitly state the relationships and data which support them. The table on p.101 contains some examples of a year-long biology sequence and some questions that relate the biology content with evolutionary theory.

Dealing with Conflicts

Evolutionary theory offers students an entry into the world of biology and it also allows students to therefore differentiate between biological and social constructs. Herein lies the biggest misunderstandings about evolution—in humans, what is biological (natural) versus what is socially prescribed (and therefore, not natural). To begin, it is pivotal for students to define the terminology and concepts required to understand evolution. For instance, students need to know that evolution is "the change in genetic frequency in a population over time"—in other words, evolution does not happen to individuals (but rather to populations), and that evolution does not have a set culminating goal. With this in mind, students begin to appreciate that in terms of evolution, all currently living organisms are equally successful. More importantly, students must appreciate that human variation is essential for the survival of our species and that these phenotypical variations are to be respected and valued. Racism and xenophobia are not biologically viable concepts, and therefore, should be abandoned.

13 A Compassionate Worldview

Patti Howell, EdD
Georgia

I live in a small, south Georgia town and work primarily as a chemistry and physics teacher. I am certified in chemistry, physics, and biology. I teach high school and I work part time as an adjunct chemistry teacher for a small college. My master's, however, is in secondary education—biology. When I applied for a job at a different high school, I was asked to teach biology. I look like a great biology teacher on paper. In reality, I have much more chemistry education and experience. For seventeen years, I taught physical sciences. When hired to teach biology, I knew I needed help. I turned to the Teacher Institute for Evolutionary Science (TIES) for training. Not only did I receive training, I also became involved in this organization. This affiliation allowed me to grow professionally and personally. In this chapter, I share my personal journey as a high school biology teacher. At the age of forty-nine, I thought I'd navigated all the significant education-related turning points. Unbeknownst to me, I'd embarked on an expedition that would take me into the murky waters of cognitive dissonance. I would pass the clear blue waters of my comfort zone and question my faith. This was a painful journey for me. One may think that I am being overly dramatic in my account to entertain readers, but I am simply relaying the upheaval that I encountered in my life.

Any reader expecting to encounter facts, data, evidence, and classroom resources for teaching evolution should skip this chapter. Here one will only find turmoil I encountered from my family, friends, and myself as I learned how to effectively teach biology from an evolutionary perspective. I will share reasons why I feel teaching evolution accurately is important and where I found the resources to teach evolution accurately.

To teach evolution, I had to learn about evolution. With all the science classes I have under my belt, I did not have a clear understanding of evolution. One thing a teacher lacks is time. Teaching any concept takes class time. It also takes something that holds more value—planning time. Most of all, it hoards the most coveted resource—personal time. Choosing to teach a concept with due diligence requires not only learning it but also keeping up with the changes unearthed by researchers eager to publish new books, journals, and dissertations.

The theory of evolution is a living, growing web that ensnares biologists, anthropologists, chemists, physicists, paleontologists—well, mostly every other "ologist" in existence. Ironically, the development of modern technology is the crucial tool to investigate archaic evidence. To understand the new information that is published daily about evolution, a teacher must continuously devour the most current literature. When a teacher commits to present a concept thoroughly and accurately, he or she is committing a lot of time to a lot of reading. To state this conundrum simply, I had to forego my nightly winding-down ritual of reading smutty romance novels and read a lot of literature about evolution to step up to

A topic in Georgia Standards of Excellence (GSE) biology standards for evolution unit is antibiotic resistance. My favorite video is *The Evolution of Bacteria on a "Mega-Plate" Petri Dish.* It can be found online in the TIES resource page. This is a short, powerful video. A large petri dish with bacteria at both ends is filmed over a period of time. As the bacteria grow to different places on the petri dish, it is exposed to antibiotics. Students see most of the bacteria die as it touches the antibiotic. They also witness the few resistant bacteria that do not die grow exponentially. At the end of the video, a phylogenic tree is placed on top of the bacteria strains. Students can see evolved bacteria species emerge! I also show an instructional video produced by the Centers for Disease Control about hand washing procedures. Since this topic coincides with flu season, students' interest peaks. This also gives me a chance to review concepts like phylogeny, classification, cellular metabolism, and characteristics of life.

One of the most important habits I must master as an educator is the correct use of lexicon. Students hear a multitude of words that are not used accurately in lay conversations. Among those words are belief, theory, and law. I recommend a video that is referenced on the TIES resource page, *Fact vs. Theory vs. Hypothesis vs. Law Explained.* This video explains that the word theory in science is used to describe why a phenomenon, like evolution, occurs. A theory, unlike a hypothesis, has been tested thoroughly and is supported by multiple lines of evidence. A belief, however, is linked to culture and values as opposed to multiple lines of evidence. A good video to explain the definition of belief is *Symbols, Values, and Norms: Crash Course Sociology #10.*

the responsibility of teaching biology effectively. Other science concepts, such as balancing chemical equations and projectile motion, are not nearly as needy. They demand only that I learn how to do them. I do not have to pay attention to those concepts anymore. The knowledge is tidily organized in my brain patiently awaiting retrieval without needing updates.

I turned to TIES for help. TIES is a well-organized, frequently updated, and peer-reviewed resource for teachers of evolution. I found many resources to inform me and keep me abreast of current research. This brings me to the aforementioned time requirement. I spent a lot of time learning how to use TIES resources. I asked myself the question, "Is it worth spending so much personal time learning about evolution? Is it worth the precious class time? I have so many standards to cover." My

answer is yes. After many careful, introspective contemplations, I now hold the conviction that this is one of the most important topics science teachers can address in the classroom.

In some ways learning about evolution is akin to learning about algebra, or studying the Bible. It is true that many people live full, happy lives without consciously using algebra, studying Biblical scriptures, or understanding evolution. With algebra, there is a hidden reward. Algebra teaches us how to think logically. Learning scripture from the Bible provides context for our moral code. What can learning evolution do to help someone?

Frequently, when I broach the subject of evolution with coworkers and friends, they preface their opening statements with the standard disclaimer "I don't believe in it, but . . ." More people use this preface in my social circles than not. That is a conversation-stopper for me. I open conversations with that phrase when I am talking about things that I find fundamentally disagreeable. For example, "I don't believe in spending all my money on lottery tickets, but, it does give a person a chance to dream big!" As soon as I hear that standard disclaimer, I think that the people are giving me a warning that they don't want to dance too close to the conversation pit leading to hell. Instead of using this as an opportunity to tell them that science is not a belief system and speaking about scientific evidence will not result in a one way ticket to eternal damnation, I shut down. I begin to wonder if there is something wrong with me. Am I just too liberal? I found the concept of evolution impossible to teach until I reconciled my belief system with my teaching practice. This led to cognitive dissonance.

I was raised as a Southern Baptist and married into a Southern Baptist family. While I occasionally visit other churches, I have always been a member of Southern Baptist churches. Southern Baptists emphasize the literal interpretation of the Bible. God made the Earth in six days. He rested on the seventh. He made Adam from the dust and Eve from his rib. From the Earth's creation, the Bible depicts a world similar to our modern Earth. Scientists, however provide a plethora of evidence that is contradictory to a literal interpretation of Genesis. In spite of that evidence, Southern Baptists desperately cling to the literal interpretation.

A strong advocate of perpetuating the bending of scientific evidence in ways that defy natural laws, Ken Ham, built the creation museum which many of my family and friends encourage me to visit. I just smile and say maybe someday. This Southern Baptist influence made my journey to evolution enlightenment downright uncomfortable. I had to face my belief system which has deep roots in my family and my identity. Not only do I hear my family and friends telling me that if I don't believe the literal interpretations of the Bible I am not a true Christian, I also see my grandparents rolling over in their graves every time I mentally deny that the Earth as we know it was created in six days. Trust me, somewhere in Kentucky there are two graves with fresh dirt. It took courage for me to even write those words on paper. I am worried that a hand may even be sticking out of one of those graves now!

I attend a Sunday school class in which my friends make fun of people who "believe" in the Big Bang Theory and the Theory of Evolution. Our congregation tends to support attempts to embed Christian Biblical teachings and principles into science curriculum which is in violation of the first amendment. I have struggled with my beliefs, worried about my salvation, and feared that Satan is stoking the fires of hell as he eagerly awaits my arrival all because I cannot accept the literal interpretation of Genesis. Worse yet, I have no one to talk to except my mentor from TIES, Bertha Vásquez and one of my too liberal for South Georgia sons, Chris. There is too much evidence that the Earth is older than six thousand years and that life began from single-celled organisms for me to see the literal interpretation of Genesis as feasible.

If one managed to read this far into my voyage without jumping ship, one may wonder where this voyage has taken me. Meandering and deviant from a straight and narrow path, I have filled my vessel with resources that I carry into my classroom confidently. Working with TIES gave me resources to teach evolution without avoiding questions and choosing only the fluffy activities that make people happy. I have the joy of stretching students' zones of proximal development more fully than before.

One reason why I feel it is important to teach evolution appropriately is that it provides a foundation for science literacy. Science literacy is a necessity in today's global society. Understanding that change is inevitable is the first step in determining what our role could and should be in its augmentation. We study world history and American history with the attitude that we will learn from our successes and failures. Doesn't studying the Earth's record of how organisms change in response to pressures help us to understand how pressures may change the biosphere for our children?

As a high school science teacher, I get the pleasure of listening to well-intended, yet bastardized explanations about how things work from people with Ph.D's from Facebook. Both adults and youth have attempted to enlighten me with their knowledge on the "climate change hoax," scary and dangerous chemicals, and how the government has a cure for AIDS that it will not share. People like to share their knowledge, or more accurately, their lack thereof. While I have the ability to listen politely as the information flows in one ear and out the other, young people have not developed mental sewage pipelines to evacuate verbal atrocities. Some of the information that youth glean from adults they respect is riddled with inaccuracies. Some information is completely false. Teachers have the burden of cleaning up misconceptions and replacing them with facts

We benefit from studying evolution. We benefit from technology used to predict storm systems so people can evacuate early. Studying evolution is no less important. Viruses and bacteria evolve to be more successful whether or not we learn why. Insects, birds, and bats will change migration patterns affecting changes in plants that rely on their pollination. These changes have occurred before. Evidence is recorded in numerous forms all around the Earth. We have the ability to learn about these changes and educate youth on the importance of understanding how their choices impact the rates of change.

My conviction about the dire need for a scientifically literate society may be biased because I love science. In order to gauge the accuracy of my claim that science literacy is critical to understanding our world, I opened the Washington Post internet page on December 26, 2018. Almost every article required a knowledge of science. A few of the articles included topics like: right to die laws, security of data chips, commercial hunting of whales, air strikes in Israel, and nuclear weapons. Even when not reading the newspaper, one makes decisions daily that require science literacy. When driving to the grocery store should one purchase fuel containing ethanol? Should one use paper or plastic grocery bags? Should one pay extra money for organic and non-GMO foods? One must have a good grasp of science literacy in order to vote. Many issues in the political arena are rooted in science. Some examples include: climate change legislation; energy production and consumption; and health care.

Unfortunately, one does not spring forth from the womb possessing science literacy. One ethical responsibility of science teachers is to provide students with the ability to set aside personal biases and distinguish the difference between data collected with due diligence from data that has been cherry-picked from special interest groups. Students should be able to make personal and political decisions based on issues—not affiliations.

As emotional, passionate, and irrational humans, we are susceptible to pseudoscience. As skeptical and rational as I try to be, I catch myself willing to throw my money into miracle cures and fad diets. Through my personal journey, I have seen evidence that so many of my beliefs are rooted in what I've been told by family, friends, and preachers. These loving, well-meaning people do not understand science issues. This is like taking career advice from someone who never held down a job.

The second reason I feel it is important to teach evolution appropriately is that common people are critical elements in either the building of a grand future or the decimation of beauty and free will through isolating ourselves from nature and each other. I know it is necessary to teach biology from an evolutionary perspective because it is a foundational theme of biology. But it was not until I embarked on this journey that I experienced resistence to the teaching of evolution. How can our society overcome its past and move forward to a productive future if we do not see that we are all a part of an intricate, colorful, related web?

There was another noteworthy event that influenced me immensely to teach biology from an evolutionary perspective. It was a strange encounter with a man in the woods. I love to hike alone. People frequently scold me for this. They say that I am strange and the woods are dangerous. To many, forests are scary places where axe murderers hide. To me, forests are much safer than my downtown south-Georgia city. I have a greater fear of people than I do nature. I was in Alabama preparing to present for TIES at the Alabama Science Teachers Association conference. While hiking, I was distracted by a cave entrance that was not on the main trail. Curiously, I wandered over to a smaller side trail. Near the cave was an older man working on a small bridge over the creek. I said hello so that I would not scare him as I crossed the bridge. We became involved in an intellectual conversation about God, the environment, and our criminal justice system. It turned out that he was a semi-retired criminal defense attorney who defended death penalty cases. He too expressed struggles with predominate religious doctrines in the South. Our conversation probably lasted over an hour. As I continued on my hike, I reflected on how my influence in the classroom could shape young minds.

BOOK RECOMMENDATIONS

There are several books that helped me to understand evolution and how to teach it effectively. A few of them that stand out as being most helpful include *The Missing Link: An Inquiry Approach for Teaching All Students About Evolution* by Lee Meadows; *The Language of God,* by Francis Collins; and *Evolution in Perspective—The Science Teacher's Compendium* from NSTA Press.

I attended a session by Lee Meadows at the Alabama Science Teachers' Conference and read his book, *The Missing Link: An Inquiry Approach for Teaching All Students About Evolution.* Seeing him as a confident presenter with knowledge and experiences to which I pale in comparison, I did not think he would have struggled with evolution education. He actually saw Lucy (the Ethiopian fossil that provides evidence of human and primate evolution from common ancestors)! Between his presentation and reading his book, I learned that he too struggled with faith and the preponderance of scientific evidence supporting the theory of evolution. In *The Missing Link* provides a framework for using inquiry methods. He contends that addressing evolution via inquiry in a non-threatening manner allows students to achieve critical science literacy to understand current issues like antiviral resistance. The reason I recommend *The Missing Link,* is that reading this and incorporating some of Meadow's techniques, I felt a little more secure in my knowledge and abilities by employing his framework for inquiry on evolution.

There is one book stands out in my mind as giving me peace. *The Language of God,* written by the geneticist who led the Human Genome Project, Francis Collins, was a revelation to me. Francis Collins is a Christian with more knowledge of genetics and evolution than most scientists. He explained why alternative theories proposed by those with religious agendas like intelligent design are flawed and riddled with statements inconsistent with evidence. It was while I read this book, at fifty years old, that I understood the meaning of **my** faith. My faith is my deep-down belief in God and my salvation through Jesus Christ. I stopped feeling like I had to match my beliefs to my Southern Baptist upbringing. I allowed myself to explore books and articles about evolutionary research without feeling anxiety. This was a profound, life altering event for me. I felt an immediate freedom and relief that I never felt before.

I felt extreme freedom and relief because I confronted personal resistance waged by my own cognitive dissonance. Confrontation, however, comes at evolution educators on multiple fronts. I had to confront myself in order to manage resistance from others. It seems that resistance is everywhere. People also get personal. I deal with resistance differently depending on the source. I follow one guideline, however, regardless of the source. I do not debate evolution. Roger Bybee is one of my favorite educational researchers. In my favorite book on teaching evolution, he penned a chapter entitled Don't Debate, Educate! Bybee also edited the book from NSTA Press, *Evolution in Perspective—The Science Teacher's Compendium.* With students, my personal responsibility is to educate. I use an activity described later in this chapter to gauge the level of resistance my students have to evolution. I focus on teaching students how to recognize and interpret reliable sources of evidence. I want to show respect and value for my students' religious beliefs. Chris Mooney and Sheril Kirshenbaum, in their book *Unscientific America—How Scientific Illiteracy Threatens Our Future,* caution against combativeness toward religion. They claim that it widens the divide unnecessarily. I see this in action when I observe debates or arguments that are deeply personal, like religion and politics. I also experience these feelings inwardly when people attack the theory of evolution like in my Sunday school class that I describe later in this chapter. I felt myself drawing away from religion until I embraced the coexistence of my religious *beliefs* and scientific evidence *acceptance.*

I posited that one of the beauties of science is that scientists from every nation, race, socioeconomic status and sexual orientation can and are encouraged to collect evidence, discuss it, debate it, and revise conclusions as more evidence is gathered. Scientists do not ridicule each other for skepticism. They do not goad into making life choices based on beliefs and biases. We all have the freedom and ability to engage in making scientifically literate decisions. We must—we owe it to our students as science teachers to empower them with science literacy. We are the only people who can.

Instead of continuing the mindset that God put all things on Earth for humans to use, I feel it is my responsibility to teach students that humans must care for people of every race and religion. It is our responsibility to care for all species of plants and animals. We are, after all, caring for our relatives. Knowing that I share common ancestry with a person allows me to see them as someone who is not so different than me. Knowing that I have a common ancestry with a lichen, a snake, and a spider makes me want to take care of our delicate environment and share its beauty with our youth. Going back to the question, "is it worth all this personal time learning about evolution so that I can teach biology effectively?"—yes. It is.

Introducing Evolution

I continuously revise the way that I teach evolution because I am exposed to new experiences with students and I learn more each year. The foundations on which I base my teaching, however, do not vary. Foremost in my heart, I want my students to become scientifically literate citizens who can understand facts and assume active roles in protecting the intricate, colorful, web of life in which we live. I also value teaching the scientific concepts in accordance with the Next Generation Science Standards. While Georgia did not adopt NGSS, Georgia Standards of Excellence are aligned with NGSS. Another core foundation of my lessons includes using local examples and experiences in addition to *textbook* examples. I rely on inquiry activities to understand students' conceptions. Finally, it is important to me that students' respect and give value to religious diversity.

There are some lessons I use each year because they align to these foundations and help students gain a deeper understanding of the mechanisms of evolution. Most of these can be found on the TIES resource web page. I teach biology from an evolutionary perspective making references to relationships from the beginning of my class, but I like to teach genetics before I teach a formal evolution unit. Throughout the year, I will emphasize key points to build up to the evolution unit. For example, while learning about organelles, we touch on the endosymbiotic theory. While learning about photosynthesis and respiration, I continuously point out that all living organisms use chemical energy. During genetics lessons, I have my students recite almost daily that all organisms' DNA have the same four nucleotides. We use anchoring phenomenon like hyperplasia in cows and humans so that students can see that humans have mechanisms and mutations in common with other organisms.

It is important for students to feel a connection to evolution that is local. In south Georgia, we have cypress swamps. Many of my students drive by them daily. As a bellringer, I assign the online article, *Cypress Trees Saw Rupturing of Earth's Supercontinents*. This article provides a tangible example of evolution that can be seen in our community. After reading this article, we have small group discussions to generate questions. After five minutes, we conduct a whole group discussion to discuss groups' questions.

 At the beginning of the formal evolution unit, I use the TIES resource "Evolution: Fact or Fiction" as an opening activity. I use it as both a warm-up activity which takes about 5 minutes and later as a group discussion probe. I extend the activity from the way it is given on the TIES resource page to incorporate NGSS elements of modeling, patterns, and analyzing data. Students are given instructions to answer the questions honestly and silently, as they will be given time to discuss them in depth later. When students complete the fact or fiction warm-up, I instruct them to go to stations. Each station contains one of the questions, pink pony beads, blue pony beads, a yes jar, and a no jar. Without discussion, female students put a pink pony bead in either the yes or no jar while males put a blue pony bead in either jar. I teach multiple biology classes each day. Students are not allowed to discuss the questions until the following day. At the end of the day, I count how many girls put pony beads in each yes or no jar vs boys. I show graphs (mathematical models) of each question as we discuss students' answers.

Undoubtedly, students will have strong reactions to these questions. I use small group discussions and whole group discussions to understand students' conceptions. I am careful to avoid revealing answers until we are finished because I want to identify prior conceptions, observe class dynamics, and establish respectfulness as a group norm. In addition, we look for patterns and differences between males' and females' conceptions. Notice I am not using the word beliefs. I purposefully use the word conceptions because as we go over the answers, I emphasize the importance of using lexicon correctly. I discuss the way we habitually misuse words like believe, theory, facts, and evidence. In short, I take a five-minute warm-up and extend it to two days. I justify the use of time as application of constructivism learning theory.

Constructivism is the concept that people construct different perceptions based on their prior knowledge and experiences (Snowman & Biehler, 2000). Evolution is a topic on which students have prior knowledge from church, family, and childhood stories about creation. The National Research Council (2000) states that one important aspect of learning is building new knowledge and understanding on what one already knows and believes. Constructivism explains how students can develop and maintain alternative conceptions. These alternative conceptions interfere with student learning (Chapman & King, 2005; Hodson, 1998; Munsun, 1994; National Research Council, 2000; Sandoval, 1995; Spencer, 2006; Smith, Disessa, & Roschelle, 1993). Research shows that in order for students to learn, instruction should identify and confront alternative conceptions. (Smith, DeSessa, and Roschelle (1993). Given the preponderance of evidence that shows the importance of understanding what students think to be able to effectively help them learn, I embrace the opportunity to spend two days on the Evolution: Fact or Fiction activity.

After opening the evolution unit with a better understanding of students' conceptions there is also an overall feeling of mutual respect for religious diversity. I generally feel as though I am able to get down to business and focus on the content standards. In Georgia, the high school biology standards place emphasis on speciation and evidence that supports the theory of evolution. The formal unit on evolution in my high school biology class is an opportunity for me to interact with my students and learn more about their ideology. It is exciting to see students transform as their thought processes are stretched in different directions. Students that need to question are often a source of delayed gratification for me. Vocalizing their beliefs and resistance does rock my boat at times, but each year my patience and commitment to teaching evolution effectively grows. One may find it ironic, but my faith also grows. At times I feel spiritually led to help students understand biology from an evolutionary perspective.

I like to use local environmental features as anchoring phenomenon. Our location in southwest Georgia was underwater until about twelve thousand years ago. We have fossil evidence of evolution all over our campus. We also have trees with interesting evolutionary histories. For example, we are surrounded by pecan trees that are artificially selected to resist pecan scab. My favorite warm up is *Cypress Tree Speciation*. One of the Georgia standards is to analyze and interpret data to explain patterns of biodiversity that result in speciation. This warm-up may take my students fifteen minutes. Students read a short article from Live Science explaining that cypress tree species exist on all continents except Antarctica. Their DNA sequence is reflective of the breakup of Pangaea. I like this because my students live among cypress trees. It is a tangible example of speciation on our school campus and in their own back yards.

Favorite Investigative Unit or Activity

After introducing the different types of speciation, my students perform Pony Bead Genetics Lab. This lab covers mechanisms of evolution: natural selection, genetic drift, random mutations, and artificial selection. Variations of this lab are popular and found easily on internet searches. I used to use a variation that involved Skittles. While Skittles (or any candy) is fun and works well for a lot of classes, I have three reasons for replacing Skittles with pony beads. First, pony beads are inexpensive and reusable. I use them for many modeling activities. They come in bags of five hundred of one color. I do not have to separate skittles into different colors. They can also be reused. Second, I teach in a lab. No eating is allowed. Hell will freeze over before students follow a *no eating in the lab rule* if I put Skittles in front of them. Finally, the pony beads can be linked together to show two alleles. I link them together with pink pipe cleaners for the female and blue pipe cleaners for the male. When calculating the allele frequencies, struggling students can count the number of each colored bead. Students can quickly and easily make models of multiple generations using the pony bead allele pairs. My students work on this lab over two days. They can easily store their pony bead alleles in zip lock bags without attracting mice and insects.

Dealing with Conflicts

I know it is necessary to teach biology from an evolutionary perspective because it is a foundational theme of biology. But it was not until I embarked on this journey that I noticed some clear, in-my-face reasons why I must teach evolution. An instance worthy of relaying occurred in one of my tenth-grade biology classes. I was telling students about how mitochondrial DNA comes from the mother. Our maternal ancestral lineage can be traced from different places in Africa. I was showing migration paths from a DNA analysis website. Two white boys vehemently argued that they didn't have "any black in them." This occurred in 2018. While I've been told too many times to count that "I didn't come from a monkey!" this was the first time that white teenagers were bold enough to voice such an aversion to having any kinship to African Americans. How can our society overcome its past and move forward to a productive future if we do not see that we are all a part of an intricate, colorful, related web?

I have spoken to a few parents who homeschool their children because they do not want their children exposed to *liberal education*. Evolution, to some parents, is a part of the larger liberal agenda to take their children away from conservative Christian values. I am not alone in experiencing controversy. At a conference last week, an attendee told me that one day, after teaching a lesson on evolution, she came back to her desk to find a Bible there, it was open to specific place so that she would be sure to see pertinent passages. I never received a Bible, but I have received enough lectures to grow weary of them.

Conclusion

Looking to the Future

Ask yourself, what are you most proud of about this country, the United States. My answers always include two things, our profound contributions to scientific discovery and our public education system. When public education is discussed, it is usually within a negative context. Overcrowding, underfunding, struggling students, and low test scores tend to crowd the conversation. This has not been my experience. I have been privileged to work with amazing, dedicated educators my whole career, both within my school district, Miami-Dade County Public Schools, and throughout the country. This humble book project is a reflection of my experience, and the reason for my optimism. Finishing this book will hopefully give the reader hope for the future. The teachers who have contributed to this project are only a tiny percentage of the many who strive to maintain high-quality academic science standards in our country's schools.

As stated in the introduction, while there is a 60 percent level of acceptance of human evolution among adults in the United States. The number rises to 68 percent in younger adults between ages 18 and 29. I am confident that these numbers will continue to rise.

While we have reason to be optimistic, we must be cautious. In 2008, Andrew Petto warned, "Anti-evolutionism is not a passing phenomenon, nor is it a matter of logic or integrity." This call for constant vigilance should always be heeded; the need to maintain the integrity of our nation's K12 evolution education should never be underestimated (Petto and Godfrey 2008).

Bertha Vázquez

Notes on Activities

Chapter 5: Appendix
Natural Selection Activity
Disciplinary Core Ideas

LS4.B: Natural Selection

Natural selection leads to the predominance of certain traits in a population, and the suppression of others. (MS-LS4-4)

LS4.C: Adaptation

Adaptation by natural selection acting over generations is one important process by which species change over time in response to changes in environmental conditions. Traits that support successful survival and reproduction in the new environment become more common; those that do not become less common. Thus, the distribution of traits in a population changes. (MS-LS4-6)

Performance Expectations

MS-LS4-4: Construct an explanation based on evidence that describes how genetic variations of traits in a population increase some individuals' probability of surviving and reproducing in a specific environment.

MS-LS4-6: Use mathematical representations to support explanations of how natural selection may lead to increases and decreases of specific traits in populations over time.

Descent with Modification and Tree of Life Activity
Disciplinary Core Idea

LS4.A: Evidence of Common Ancestry and Diversity

Anatomical similarities and differences between various organisms living today and between them and organisms in the fossil record, enable the reconstruction of evolutionary history and the inference of lines of evolutionary descent (MS-LS4-2)

Performance Expectation

MS-LS4-2: Apply scientific ideas to construct an explanation for the anatomical similarities and differences among modern organisms and between modern and fossil organisms to infer evolutionary relationships.

Bibliography

Introduction

"Academic Freedom" Legislation. National Center for Science Education, 2009. http://ncse.com/creationism/general/academic-freedom-legislation

"Chronology of 'Academic Freedom' Bills." National Center for Science Education, 2013. http://ncse.com/creationism/general/chronology-academic-freedom-bills

Edwards v. Aguillard, 482 U.S. 578 (1987)

Epperson v. Arkansas, 393 U.S. 97 (1968)

Kitzmiller v Dover, 400 F. Supp. Page 137 (M.D. Pa. 2005).

Petto, Andrew and Laura R. Godfrey. *Scientists Confront Intelligent Design and Creationism*. New York: W.W. Norton, 2008.

Scott, Eugenie C. *Evolution vs. Creationism: An Introduction*. Berkeley & Los Angeles, California: University of California Press, 2004.

Chapter 1. Bertha Vázquez

Berkman, M., and E. Plutzer. *Evolution, Creationism and the Battle to Control America's Classrooms*. New York: Cambridge Press, 2010.

Berkman MB, Pacheco JS, Plutzer E. "Evolution and Creationism in America's Classrooms: A National Portrait." PLoS Biol 6(5) (2010): e124. doi:10.1371/journal.pbio.0060124

Branch, G., Reid, A. & Plutzer, E. Teaching Evolution in U.S. Public Middle Schools: Results of the First National Survey. Evo Edu Outreach 14, 8 (2021). https://doi.org/10.1186/s12052-021-00145-z

Carroll, Sean (2014) Is America Evolving on Evolution? Scientific American. Retrieved from https://www.scientificamerican.com/article/is-america-evolving-on-evolution/

Dawkins, R. *The Greatest Show on Earth: The Evidence for Evolution*. New York: Free Press, 2009.

Gishlick, Alan D. "Icons of Evolution?" Retrieved from https://ncse.com/creationism/analysis/icons-evolution

Griffith J, Brem S. "Teaching Evolutionary Biology: Pressures, Stress, and Coping." *J Res Sci Teach* 41 (2004): 791–809.

Shubin, N. *Your Inner Fish: A Journey into the 3.5-billion-year History of the Human Body*. New York: Pantheon Books, 2008.

Plutzer, E., Branch, G. & Reid, A. "Teaching Evolution in U.S. Public Schools: A Continuing Challenge." Evo Edu Outreach 13, 14 (2020). https://doi.org/10.1186/s12052-020-00126-8

Zombie Activity, Troup County Schools, GA. Retrieved from: http://www.troup.k12.ga.us/userfiles/929/my%20files/science/ms%20science/7th%20science/evolution/evolution_resources_schoolpointe.pdf?id=21261

Chapter 3. Amanda Clapp

Anderson, Laurie Halse. *Fever, 1793*. New York: Simon & Schuster Books for Young Readers, 2000.

The Biology of Skin Color HHMI BioInteractive http://www.hhmi.org/biointeractive/biology-skin-color

Duncan, B., and Riggs, B. *Cherokee Heritage Trails Guidebook*. Chapel Hill, NC: Museum of the Cherokee Indian by University of North Carolina Press, 2003.

Flynn, AK. "TED-Ed. The Science of Skin Color." https://www.ted.com/talks/angela_koine_flynn_the_science_of_skin_color

Jablonski, N. 2009. Nina Jablonski: Skin color is an illusion [Video File]. Retrieved from https://www.ted.com/talks/nina_jablonski_breaks_the_illusion_of_skin_color

Keeley, P. *et al. Uncovering student ideas in science*. Vol. 3. Arlington, VA: NSTA Press, 2008.

National Research Council. *A Framework for K-12 Science Education: Practices, Crosscutting Concepts, and Core Ideas*. Washington, DC: The National Academies Press, 2012. https://doi.org/10.17226/13165.

"Popped Secret: The Mysterious Origin of Corn." HHMI. Retrieved October 4, 2018, https://www.hhmi.org/biointeractive/popped-secret-mysterious-origin-corn.

Schrein, Caitlin. Quote retrieved from the NCSE blog by Stephanie Keep, "America's Unwillingness to Accept Evolution En Masse is . . . Complicated" (May 2016)

Shubin, N. *Your Inner Fish: A Journey into the 3.5-billion-year History of the Human Body*. New York: Pantheon Books, 2008.

Chapter 4. Kenny Coogan

Banner, Jr., J. M., & Cannon, H. C. *The Elements of Teaching*. New Haven, CT: Yale University Press, 1997.

Barnes, M. E., & Brownell, S. E. "A Call to Use Cultural Competence When Teaching Evolution to Religious College Students: Introducing Religious Cultural Competence in Evolution Education" (ReCCEE). *CBE—Life Science Education* 16, no. 4 (2017): 1–10. https://doi.org/10.1187/cbe.17-04-0062

Barnes, M. E., Elser, J. & Brownell, S. E. "Impact of a Short Evolution Module on Students' Perceived Conflict Between Religion and Evolution." *The American Biology Teacher* 79, no. 2 (2017): 104–111.

Clary, R. "Diffusing Discomfort: Bridging Philosophical and Religious Conflicts through Reflective Writing," *The Science Teacher* 84, no. 2 (2017): 26–30.

Collins, F. *The Language of God: A Scientist Presents Evidence for Belief*. New York: Free Press, 2006.

Cooper, R. A. "The Goal of Evolution Instruction: Should we Aim for Belief or Scientific Literacy?" *Reports of the National Center for Science Education* 21, no. 1-2 (2001): 18–18.

Cooper, R. A. "Scientific Knowledge of the Past is Possible: Confronting Myths about Evolution and Scientific Methods." *The American Biology Teacher* 64, no. 6 (2002): 427–432.

Cooper, R. A. "How Evolutionary Biologists Reconstruct History:

Patterns and Processes." *The American Biology Teacher* 66, no. 2 (2004): 101–108.

Cooper, R. A. (2017). *Evolution as a Theme in Biology.* Unpublished manuscript. Available at http://bit.ly/EvolAsATheme.

Dawkins, R. *The God Delusion.* London: Bantam Press, 2006.

Dobzhansky, T. *The Biology of Ultimate Concern.* New York: World Publishing Company, 1969.

Dobzhansky, T. "Nothing in Biology Makes Sense Except in the Light of Evolution." *The American Biology Teacher* 35, no. 3 (1973): 125–129.

"Evolution: Fossil, genes, and mousetraps." HHMI BioInteractive. 2007. https://www.hhmi.org/biointeractive/fossils-genes-and-mousetraps.

Gould, S. J. *Rock of Ages: Science and Religion in the Fullness of Life.* New York: Ballantine, 1999.

Jacob, F. (1977). "Evolution and tinkering." *Science, 196,* 1161–1166.

Lucci, K., & Cooper, R.A. (In press). "Using the I² strategy to help students think like biologists about natural selection." *The American Biology Teacher.*

Meagher, T. R., & Futuyma, D. J. (2001). "Evolution, Science, and Society." *The American Naturalist* 158, no. S4: S1–S46. https://doi.org/10.1086/509090. Accessed at http://www.evolutionsociety.org/file.php?file=sitefiles/Evo_Statement_1997_775_kb.pdf.

Miller, K. R. *Finding Darwin's God: A Scientist's Search for Common Ground between God and Evolution.* New York: HarperTrade, 1999.

National Academy of Sciences. *Teaching about Evolution and the Nature of Science.* Washington, DC: National Academy Press, 1998. Accessed at https://www.nap.edu/catalog/5787/teaching-about-evolution-and-the-nature-of-science.

National Institute of Neurological Disorders and Stroke. *Low back pain fact sheet.* (2018, July) Accessed at https://www.ninds.nih.gov/Disorders/Patient-Caregiver-Education/Fact-Sheets/Low-Back-Pain-Fact-Sheet.

Nelson, C. "Is Evolution Weak Science, Good Science, or Great Science?" In *The Creation Controversy & and the Science Classroom:* 19–50. Arlington, VA: The NSTA Press, 2000a.

Nelson, C. (2000b). "Is Evolution Weak Science, Good Science, or Great Science?" Accessed at http://www.indiana.edu/~ensiweb/teach.fs.html.

Nesse, R. M., & Williams, G. C. *Why We Get Sick: The NewScience of Darwinian Medicine.* New York: Random House, 1994.

Olshansky, S. J., Carnes, B. A., Butler, R. N. "If Humans Were Built to Last." *Scientific American,* 284, no. 3 (March 2001): 50–55.

Roughgarden, J. *Evolution and Christian faith: Reflections of an evolutionary biologist.* Washington, DC: Island Press, 2006.

Sager, C., (Ed.). *Voices for Evolution,* 3rd ed. Berkeley (CA): National Center for Science Education, 2008. Accessed at http://ncse.com/voices.

Schwab, J.J. *The Teaching of Science as Inquiry.* In J.J. Schwab, & P.F. Brandwein, (Eds.), *The Teaching of Science* (pp. 1–103). Cambridge, MA: Harvard University Press, 1962.

Scott, E. C. (2009). "The Creation/evolution Continuum." National Center for Science Education. Accessed at https://ncse.com/library-resource/creationevolution-continuum.

Shubin, N. *Your Inner Fish: A Journey into the 3.5-billion-year History of the Human Body.* New York: Pantheon Books, 2008.

Smith, M. U. (1994). Counterpoint: Belief, Understanding, and the Teaching of Evolution. *Journal of Research in Science Teaching, 31,* 591–597.

Truong, J. M., Barnes, M. E., & Brownell, S. E.. Can Six Minutes of Culturally Competent Evolution Education Reduce Students' Level of Perceived Conflict Between Evolution and Religion? *The American Biology Teacher 80,* no. 2 (2018): 106–115.

Westerling, K. (1992). *Baggie cladistics.* Accessed at http://www.indiana.edu/~ensiweb/lessons/clad.bag.ms.pdf. (Background information available at http://www.indiana.edu/~ensiweb/lessons/clad.bag.html.)

Woods, C. S., & Scharmann, L. C. "High School Students' Perceptions of Evolutionary Theory." *Electronic Journal of Science Education 6, no.* 2. (2001) Accessed at https://wolfweb.unr.edu/homepage/crowther/ejse/woodsetal.html.

Chapter 5. Robert A. Cooper

"Are Birds Really Dinosaurs?" University of California Museum of Paleontology (2020). Retrieved January 18, 2020. https://ucmp.berkeley.edu/diapsids/avians.html.

Bloom, P., & Weisberg, D.S. "Childhood Origins of Adult Resistance to Science." *Science* 316 (2007): 996–997.

Brusatte, S. *The Rise and the Fall of the Dinosaurs: The Untold Story of a Lost World.* New York: Macmillan, 2018.

Carroll, S.B. *The Making of the Fittest: DNA and the Ultimate Forensic Record of Evolution.* New York: W. W. Norton, 2006.

Cooper, R.A. "How Evolutionary Biologists Reconstruct History: Patterns and Processes." *The American Biology Teacher* 66, no. 2 (2004): 101–108.

Cooper, R.A. "Natural Selection as an Emergent Process: Instructional Implications." *Journal of Biological Education* 51, no. 3 (2017): 247–260.

Cooper, R. A. Scientific Knowledge of the Past is Possible: Confronting Myths about Evolution and Scientific Methods. *The American Biology Teacher* 64, no. 6 (2002): 427–432.

Darwin, C. *On the Origin of Species by Means of Natural Selection.* Cambridge, MA: Harvard University Press, 1964.

Gould, S.J. *Wonderful Life: The Burgess Shale and the Nature of History.* New York: W. W. Norton, 1989.

Grant, B.R., & Grant, P.R. "What Darwin's Finches Can Teach Us About the Evolutionary Origin and Regulation of Biodiversity," *BioScience* 53, no. 10 (2003): 965–975. http://bioscience.oxfordjournals.org/content/53/10/965.full.pdf+html.

"Great Transitions: The Origin of Humans." HHMI BioInteractive (2014a). Retrieved January 18, 2020 from https://www.biointeractive.org/classroom-resources/great-transitions-origin-humans.

"Great Transitions: The Origin of Tetrapods." HHMI BioInteractive (2014b). Retrieved January 18, 2020 from https://www.biointeractive.org/classroom-resources/great-transitions-origin-tetrapods.

"Great Transitions: The Origin of Birds." HHMI BioInteractive. (2015). Retrieved January 18, 2020 from https://www.biointer-active.org/classroom-resources/great-transitions-origin-birds.

Halverson, K.L. "Using Pipe Cleaners to Bring the Tree of Life to Life." *The American Biology Teacher* 72, no. 4 (2010): 223–224.

"How does Natural Selection Work?" American Museum of Natural History. January 18, 2020, https://www.amnh.org/exhibitions/darwin/evolution-today/how-does-natural-selection-work.

Jacob, F. "Evolution and Tinkering." *Science*, 196 (1977): 1161–1166.

Lents, N.H. *Human Errors: A Panorama of our Glitches, From Pointless Bones to Broken Genes.* Boston: Houghton Mifflin Harcourt, 2018.

Lucci, K., & Cooper, R.A. "Using the I² Strategy to Help Students Think Like Biologists about Natural Selection." *The American Biology Teacher* (2019).

Marcus, G. *Kluge: The Haphazard Construction of the Human Mind.* Boston: Houghton Mifflin, 2008.

Mayr, E. *One Long Argument: Charles Darwin and the Genesis of Modern Evolutionary Thought.* Cambridge, MA: Harvard University Press, 1991.

National Institute of Neurological Disorders and Stroke (2018, July). Low back pain fact sheet. Retrieved January 18, 2020 from https://www.ninds.nih.gov/Disorders/Patient-Caregiver-Education/Fact-Sheets/Low-Back-Pain-Fact-Sheet.

Nesse, R. M., & Williams, G. C. *Why We Get Sick: The New Science of Darwinian Medicine.* New York: Random House, 1994.

NGSS Lead States. *Next generation science standards: For states, by states.* Washington, DC: National Academies Press, 2013.

Olshansky, S. J., Carnes, B. A., Butler, R. N. "If Humans Were Built to Last." *Scientific American* 284, no. 3 (March 2001): 50–55.

"The Origin of Species: The Beak of the Finch." HHMI BioInteractive (2013). Retrieved January 18, 2020 from https://www.hhmi.org/biointeractive/origin-species-beakfinch.

Shtulman, A. *Scienceblind: Why our intuitive theories about the world are so wrong.* New York, NY: Basic Books, 2017.

Shubin, N. *Your Inner Fish: A Journey into the 3.5-billion-year History of the Human Body.* New York: Pantheon Books, 2008.

Sinatra, G.M., Brem, S.K., & Evans, E.M. "Changing minds? Implications of Conceptual Change for Teaching and Learning about Biological Evolution." *Evolution: Education and Outreach* 1 (2008): 189–195.

Weiner, J. *The Beak of the Finch: A Story of Evolution in our Time.* New York: Alfred A. Knopf, 1994.

"What did *T. rex* taste like?" University of California Museum of Paleontology. (2020b). Retrieved January 18, 2020 from https://ucmp.berkeley.edu/education/explorations/tours/Trex/index.html.

Chapter 7. Reginald Finley, Sr.

Fairbanks, D. J. *Relics of Eden: the Powerful Evidence of Evolution in Human DNA.* Buffalo, NY: Prometheus Books, 2009.

The Secular Web at: https://infidels.org/

Chapter 8. Katie Green

BioInteractive. 2018. The Howard Hughes Medical Institute. September 30, 2018. https://www.hhmi.org/biointeractive

"Broader Social Impacts Committee." *Human Evolution.* 2018. The Smithsonian Institution's Human Origins Program. September 30, 2018 https://humanorigins.si.edu/about/broader-social-impacts-committee.

"Evolution 101." *Understanding Evolution.* 2018 University of California Museum of Paleontology. September 30, 2018. https://evolution.berkeley.edu/evolibrary/article/evo_01.

"Evolution: Fossils, Genes and Mousetraps." The Howard Hughes Medical Institute. September 30, 2018. https://www.hhmi.org/order-materials/holiday-lecture/evolution-fossils-genes-and-mousetraps

"The History of Evolutionary Thought." Understanding Evolution. University of California Museum of Paleontology. September 30, 2018. https://evolution.berkeley.edu/evolibrary/article/0_0_0/history_01

"Human Evolution." The Smithsonian Institution's Human Origins Program. September 30, 2018. https://humanorigins.si.edu/

"The Origin of Species: The Making of a Theory." BioInteractive. 2013. The Howard Hughes Medical Institute. September 30, 2018. https://www.hhmi.org/biointeractive/origin-species-making-theory

"Science at Multiple Levels." Understanding Science. University of California Museum of Paleontology. September 30, 2018 https://undsci.berkeley.edu/article/howscienceworks_19

"Understanding Evolution." University of California Museum of Paleontology. September 30, 2018. https://evolution.berkeley.edu/evolibrary/home.php

"Understanding Science." 2018. University of California Museum of Paleontology. September 30, 2018. http://www.understandingscience.org

Chapter 10. David Mowry

BioInteractive. 2018. The Howard Hughes Medical Institute. September 30, 2018. https://www.hhmi.org/biointeractive

"Broader Social Impacts Committee." Human Evolution. 2018. The Smithsonian Institution's Human Origins Program. September 30, 2018. https://humanorigins.si.edu/about/broader-social-impacts-committee

"Evolution 101." Understanding Evolution. 2018 University of California Museum of Paleontology. September 30, 2018. https://evolution.berkeley.edu/evolibrary/article/evo_01.

"Evolution: Fossils, Genes and Mousetraps." 2006. The Howard Hughes Medical Institute. September 30, 2018. https://www.hhmi.org/order-materials/holiday-lecture/evolution-fossils-genes-and-mousetraps

"The History of Evolutionary Thought." Understanding Evolution. 2018. University of California Museum of Paleontology. September 30, 2018. https://evolution.berkeley.edu/evolibrary/article/0_0_0/history_01

"Human Evolution." 2018. The Smithsonian Institution's

Human Origins Program. September 30, 2018. https://humanorigins.si.edu/

"The Origin of Species: The Making of a Theory." BioInteractive. The Howard Hughes Medical Institute. September 30, 2018. https://www.hhmi.org/biointeractive/origin-species-making-theory

"Science at Multiple Levels." Understanding Science. University of California Museum of Paleontology. September 30, 2018 https://undsci.berkeley.edu/article/howscienceworks_19

"Understanding Evolution." University of California Museum of Paleontology. September 30, 2018. https://evolution.berkeley.edu/evolibrary/home.php

"Understanding Science." University of California Museum of Paleontology. September 30, 2018. http://www.understandingscience.org

Chapter 11. Blake Touchet

Achenbach, J. "The people who saw evolution." *Princeton Alumni Weekly* (April 23, 2014). Retrieved September 22, 2018, from https://paw.princeton.edu/article/people-who-saw-evolution

"Evolution: Online Course for Teachers: Session 1 What is the Nature of Science?" (2001). Retrieved September 22, 2018, from http://www.pbs.org/wgbh/evolution/educators/course/session1/

Funk, C. (2015, January 29). "Public and Scientists' Views on Science and Society." *Pew Research Center*. Retrieved from http://www.pewinternet.org/2015/01/29/public-and-scientists-views-on-science-and-society/

Lederman, N. G. "Students' and Teachers' Conceptions of the Nature of Science: A Review of the Research." *Journal of Research in Science Teaching* 29, no. 4 (1992): 331-359. doi:10.1002/tea.3660290404

McKinney, D. & Michalovic, M. "Teaching the Stories of Scientists and their Discoveries." *NSTA WebNews Digest.* (October 29, 2004). Retrieved from http://www.nsta.org/publications/news/story.aspx?id=49940

Musante, S. "Learning the nature of science." *BioScience* 55, no. 10 (2005): 833–833. doi:10.1641/0006-3568(2005)055[0833:ltnos]2.0.co;2

Narguizian, P. "Understanding the Nature of Science through Evolution. *NSTA WebNews Digest.* (October 29, 2004) Retrieved from http://www.nsta.org/publications/news/story.aspx?id=49939

National Center for Science Education. (2018). Retrieved September 22, 2018. https://ncse.com/

O'Malley, M. A. "Endosymbiosis and its Implications for Evolutionary Theory." *Proceedings of the National Academy*

of Sciences 112, no. 33 (2015): 10270–10277. https://doi.org/10.1073/pnas.1421389112

"Resources for Teaching and Learning Biology." Retrieved September 22, 2018. https://www.aibs.org/education/teaching_resources.html

Teacher Institute for Evolutionary Science. Retrieved September 22, 2018. https://www.richarddawkins.net/ties/

van Wyhe, J. "The Complete Work of Charles Darwin Online." (2002). Retrieved from http://darwin-online.org.uk

van Wyhe, J. "Wallace Online." (2012). Retrieved from http://wallace-online.org/

Watson, James, Francis Crick, Maurice Wilkins, and Rosalind Franklin. (February 23, 2018). Retrieved September 22, 2018, from https://www.sciencehistory.org/historical-profile/james-watson-francis-crick-maurice-wilkins-and-rosalind-franklin

Chapter 12. David Upegui

Dobzhansky, T. (1973). "Nothing in Biology Makes Sense except in the Light of Evolution." *The American Biology Teacher* 35, no. 3: 125-129. doi:10.2307/4444260

Jablonski, N. G., & Chaplin, G. "Human skin pigmentation as an adaptation to UV radiation." *Proceedings of the National Academy of Sciences* 107, no. 2 (2010): 8962-8968.

Henrich, J. P. *The Secret of our Success: How Culture is Driving Human Evolution, Domesticating our Species, and Making us Smarter.* Princeton, NJ: Princeton Univ. Press, 2016.

Vance, M. *Biology Teaching in a Racist Society.* In D. Gill & L. Levidow (Eds.), *Anti-Racist Science Teaching* (pp. 107-123). London: Free Association Books, 1987.

Wilson, E. O. *The Social Conquest of Earth.* New York: Liveright Pub, 2012.

Chapter 13. Patti Howell

Fact vs. Theory vs. Hypothesis vs. Law Explained https://tieseducation.org/resource/fact-vs-theory-vs-hypothesis-vs-law-explained-itsokaytobesmart/

Parry, W. "Cypress Trees Saw Rupturing of Earth's Supercontinents." Live Science (2012). https://www.livescience.com/20109-pangaea-cypress-family-tree.html

Petri Dish Video https://tieseducation.org/resource/the-evolution-of-bacteria-on-a-mega-plate-petri-dish/

Symbols, Values, and Norms: Crash Course Sociology #10. Open Education Sociology Dictionary. https://sociologydictionary.org/belief/

TIES Preconceptions Bell ringer: https://tieseducation.org/resource/bell-ringer-1/

Index

Note: Photographs and associated captions are indicated by *f* following the page number.

A

academic freedom bills, 2
activities. *See also* bell-ringers
    beak depth of finch, 13–14, 34–38, 35*f*, 36*f*
    "Biogeography of the Malay Archipelago," 20
    Blossoms, 100
    Cypress Tree Speciation, 109
    "Evolution: Fact or Fiction," 108–109
    "Family Reunion," 43
    "Fish or Mammals?" case study, 45
    hominin craniometry lab, 72, 76, 79
    Modeling Evolution: The Charlie Shuffle, 78–79
    on natural selection, guidelines for, 112
    Oh Deer! (game), 23
    OneZoom website, 9, 54
    peppered moth simulation, 23, 28, 53
    Phylogenetic Tree of Anole Lizards, 100
    Pony Bead Genetics Lab, 109–110
    "A Rainbow of Sepia," 89–90
    rotation lab, 30–31
    School Board Scenario, 58, 59–60
    skull analysis, 60, 72, 76, 79
    Spot the Misconception (game), 88–89, 92
    "Survival of the Sneakiest" (story), 100
    "This Lab is for the Birds," 13–14
    "thumb," 100
    Time Machine (online game), 14, 54
    "What Did *T. rex* Taste Like?," 38–40
agriculture, evolution and, 5–6, 30
Aikenhead, Glen, 56–57
Alabama Science Teachers Association, 106, 107
al-Hasan, Abu Ali, 100
The Ancestor's Tale: A Pilgrimage to the Dawn of Evolution (Dawkins), 4, 9
Andersen, Paul, 98
Anderson, Laurie Halse, 26
The Annotated Origin (Costa), 25
AP biology curriculum, 98–99
Arkansas Act 590 (1981), 2
artificial selection, 5*f*, 6, 30
Atheism, Freedom, and Liberation show (AFL), 49
Australopithecus sediba fossils, 64–65
autosomal traits, 47–48

B

Bacteria Growing on a Petri Dish (video), 103
Bakker, Bob, 64

banning of evolution education, 1–2, 4
Bateson, Mary Catherine, 85
beak depth of finch (activity/video), 13–14, 34–38, 35*f*, 36*f*, 39, 65
The Beak of the Finch: A Story of Evolution in Our Time (Weiner), 34
bell-ringers. *See also* activities
    "Bones, Stones, and Genes" poster discussion, 20
    cheetah and gazelle natural selection, 44
    Cypress Trees Saw Rupturing of Earth's Supercontinents, 108
    natural selection fitness scores/adjustments, 71
    "Those Pesky Fleas," 10–11, 30
Berger, Lee, 64–70, 66*f*
Berger, Matthew, 64
Bertka, Connie, 76
"Biogeography of the Malay Archipelago" (activity), 20
Biology (Miller and Levine), 78
Biology as Ideology (Lewontin), 97
"The Biology of Skin Color," 89
Blossoms (video and activities), 100
Bone Clones Inc., 43, 60
"Bones, Stones, and Genes" (HHMI poster), 20
books
    The Ancestor's Tale (Dawkins), 4, 9
    The Annotated Origin (Costa), 25
    The Beak of the Finch (Weiner), 34
    Biology (Miller and Levine), 78
    Biology as Ideology (Lewontin), 97
    The Dinosaur Heresies (Bakker), 64
    Early Man, 61
    Evolution in Perspective—The Science Teacher's Compendium (NSTA), 107
    Evolution vs. Creationism (Scott), 1
    Evolution vs. Creationism: Inside the Controversy (Scientific American), 77
    Fever 1793 (Anderson), 26
    Ghost Stories for Darwin (Subramaniam), 97
    The Greatest Show on Earth (Dawkins), 8, 42
    How and Why Wonder Book, 15
    Human Errors (Lents), 9
    The Journey of Man (Wells), 64
    The Language of God (Collins), 107
    Lucy (Johanson), 61
    Magic of Reality (Dawkins), 19, 87
    The Missing Link (Meadows), 107
    Monkey Girl (Humes), 56
    The Nature of Race (Morning), 97
    One Long Argument (Mayr), 38

Only a Theory (Miller), 97

On the Origin of Species (Darwin), 25, 38, 86

On the Origin of Species: The Illustrated Edition (Quammen, illustrator), 25

Racism, Not Race (Graves and Goodman), 97

Relics of Eden (Fairbanks), 50

Remarkable Creatures (Carroll), 67

The Rise and the Fall of the Dinosaurs (Brusatte), 34, 39

Scientists Confront Creationism (Petto and Godfrey), 1

The Selfish Gene (Dawkins), 4

The Skull in the Rock (Berger), 64

Unnatural Selection (van Grouw), 32

Unscientific America—How Scientific Illiteracy Threatens Our Future (Mooney and Kirshenbaum), 107

The Variation of Animals and Plants under Domestication (Darwin), 32

Why Evolution is True (Coyne), 50

Why We Believe (Fuente), 97

Your Inner Fish (Shubin), 4, 9, 25, 36

Zoobiquity (Natterson-Horowitz), 18

Boswell, John, 43

Boyer, Paul, 16

brave spaces, 90–91

Brusatte, Steve, 34, 39

Burt, Cyril L., 95

Bybee, Roger, 107

C

Calabrese, Mr., 15

Calabrese-Barton, Angela, 95

Carroll, Sean B., 10, 39, 40, 67

Center for Inquiry, 9

Centers for Disease Control, 85, 104

Chambers, Nikki, 15, 19*f*

Cherokee creation story, 21, 24

chimpanzees

   Goodall's work with, 63–64

   human relationship to, 32, 40, 52, 63 (*see also* primates, human relationship to)

cladistics/cladograms, 31, 32, 38–40, 54

Clapp, Amanda, 21, 24*f*

collateral learning, 57–58

College Board, 98, 99

Collins, Francis, 107

conflicts, dealing with, 14, 20, 28, 31–32, 40, 45–46, 54, 72, 79–80, 90–91, 102, 110

constructivism, 109

Coogan, Kenny, 29

Cooper, Robert A., 33

Costa, Jim, 25

cousins, relationships as, 32, 39, 48, 52, 54

Cowles, Elizabeth, 98

Coyne, Jerry, 49, 50

CRAAP (Currency, Relevance, Authority, Accuracy, and Purpose) test, 88, 90

creationism. *See also* religious beliefs

   academic freedom bills on teaching, 2

   creation stories or myths, 21, 24, 55, 57–58

   equal time laws on teaching, 2, 45, 62

   intelligent design and, 2, 58, 60, 107

   not justifying as science, 54, 79–80

   presented as science, 2, 45–46, 58, 80

   pressure to include in curriculum, 8

   research on inclusion in curriculum, 58, 60

critical thinking skills, 26, 49–50, 53, 90, 93, 94, 96, 98

"Cultural and Religious Sensitivity Teaching Strategies Resource" (Smithsonian Museum), 20

cultural border crossings, 56–58

Cypress Tree Speciation (activity), 109

Cypress Trees Saw Rupturing of Earth's Supercontinents (bell-ringer), 108

D

Darwin, Charles

   Dawkins and, 4

   discrediting of, 5

   evolutionary predictions by, 53

   on grandeur of evolutionary viewpoint, 77

   influences on, 83

   On the Origin of Species, 25, 38, 86

   teaching theories of (*see* evolution education; natural selection)

   The Variation of Animals and Plants under Domestication, 32

   voyage and discoveries of, 23, 44, 61, 65, 75, 76

   Wallace and, 23, 42, 61, 75, 76, 83, 86

Darwin, Emma, 86

Darwin, Erasmus, 61

Darwin Day, 27, 30

Dawkins, Richard

   The Ancestor's Tale, 4, 9

   on evolution as theory, 76

   Evolution vs. Creationism contribution, 77

   The Greatest Show on Earth, 8, 42

   influence of, 4–5

   interview of, 49

   Magic of Reality, 19, 87

   natural selection debate, 2

   photograph of, 3*f*

   on science as poetry of reality, 43

   The Selfish Gene, 4

   Why are There Still Chimpanzees? (video), 40

The Dinosaur Heresies (Bakker), 64

Dobzhansky, Theodosius, 61

Duncan, Chance, 41, 45*f*

E

Early Man, 61

Edwards v. Aguillard (1987), 2

Elliott, Marina, 68
endosymbiotic theory, 86, 96, 100, 108
ENSI/SENSI, 98
Epperson v. Arkansas (1968), 1, 2, 45
equal time laws, 2, 45, 62
evidence
  claim-evidence-reasoning approach, 25, 27–28
  confidence in, 85
  CRAAP test, 88, 90
  critical thinking based on, 26, 50, 90
  evaluation of validity of, 25, 26, 59, 84–85, 88–89, 90, 107
  evolution as fact based on, 52, 76, 77, 79, 109
  science experiment multiplicity for, 59, 85, 108
"Evolution 101" webpage, 76
evolutionary trees, 39, 40, 40*f. See also* pedigree charts; phylogenetic
    trees; primates, human relationship to
evolution education
  academic freedom bills on, 2
  acceptance rates, 2, 10, 53, 60, 71, 111
  activities in (*see* activities; bell-ringers)
  banning of, 1–2, 4
  belief in, qualifying language, 8–9, 104–105
  challenges in US, 8–10, 74–77, 94–96, 98
  complexity of, teaching despite, 82, 84
  conflicts due to (*see* conflicts, dealing with)
  continued discoveries in, 64–71, 66*f*, 83, 103
  cost-effective approach to, 78*f*
  cultural border crossings with, 56–58
  curriculum for, 17–18, 22–23, 45, 47–48, 51–53, 63–64, 75–77,
    98–102 (*see also* activities; bell-ringers; *introduction of* subentry)
  discrediting of, 5, 50, 62
  embedding of, 25–26, 32, 42–43, 44, 96
  equal time laws, 2, 45, 62
  evidence in (*see* evidence)
  future of, 111
  genetic mutations, 11–14, 30, 48, 52, 71, 78 (*see also* artificial
    selection; natural selection)
  history of US, 1–2
  importance and value of, 5, 23–25, 104, 105–106
  interdisciplinary approach to, 26, 63, 99, 100–102
  introduction of, 10–13, 19–20, 26–28, 30, 34–38, 43–45, 54, 60, 71,
    77–79, 88–89, 100, 108–109
  misconceptions in, 11, 27–28, 32, 34, 40, 52–53, 56, 74, 75–77, 83,
    88–89, 92, 105–106
  nonscientists' contributions to, 70
  organism of the day in, 47, 49, 54
  origins of life not in, 52–53
  randomness discussion, 11–12, 14, 71, 78
  recommendations for, 25–26, 98–100
  relevance of, supporting, 26, 32, 53
  religious beliefs and (*see* religious beliefs)
  research on, 50–52, 53, 56–60, 61–62

  resources for (*see* books; Teacher Institute for Evolutionary
    Science; videos; web resources)
  social justice issues in, 84, 89, 94–95, 99–100, 102
  storytelling or sciencetelling in, 64, 67, 68, 71, 85–87
  student stress from conflicts over, 8, 74
  teachers of (*see* science teachers)
  teamwork and collaboration importance to, 70, 83
  terminology in, 43, 55, 74, 102, 109
  as theory, clarification of, 27–28, 32, 52, 56, 58, 59–60, 76
  3-D printing in, 69–70, 72
"Evolution: Fact or Fiction" (activity), 108–109
Evolution in Perspective—The Science Teacher's Compendium (NSTA), 107
"Evolution Myths: Evolution is just so Unlikely" *(New Scientist),* 14
The Evolution of Bacteria on a 'Mega-Plate' Petri Dish (video), 12–13
Evolution vs. Creationism (Scott), 1
*Evolution vs. Creationism: Inside the Controversy (Scientific
    American),* 77
"Evolution: Watching Speciation Occur" *(Scientific American),* 13
Excellent and Simple Explanation of Natural Selection (video), 12–13

F

Facts vs. Theory (video), 28
Fact vs. Theory vs. Hypothesis vs. Law Explained (video), 104
Fairbanks, Daniel, 50, 53
"Family Reunion" (activity), 43
Fastovksy, David, 95*f,* 96
Fever 1793 (Anderson), 26
films. *See* videos
Finley, Reginald, Sr., 47
Fisher, Ronald, 83
"Fish or Mammals?" case study, 45
Florida Association of Science Teachers, 29
Flynn, Angela, 26
Food and Drug Administration, US, 85
Fossils, Genes, and Mousetraps (video), 76–77
Frameworks for Science Education, 25
Franklin, Rosalind, 86, 100
Fuente, Agustin, 97
The Fun Scientists LLC, 50

G

genetic mutations, 11–14, 30, 48, 52, 71, 78. *See also* artificial
    selection; natural selection
genetic traits, 47–48, 54, 93, 96. *See also* skin color/pigmentation
Ghost Stories for Darwin (Subramaniam), 97
Gloucester, Mr., 57–60
Godfrey, Laura R., 1
Goodall, Jane, 63–64
Goodman, Alan H., 97
Gould, Stephen J., 62
Grant, Peter and Rosemary, 13–14, 34, 35–36, 65, 77, 86
Graves, Joseph L., Jr., 97
The Greatest Show on Earth: The Evidence for Evolution (Dawkins), 8, 42

"Great transitions: The Origin of Birds" (video), 36
"Great transitions: The Origin of Humans" (video), 36
"Great transitions: The Origin of Tetrapods" (video), 36
Green, Katie, 55, 59f
Gurche, John, 70

H

Haida creation story, 57–58
Haldane, JBS, 83
Ham, Ken, 105
handwashing (video), 104
Hardy-Weinberg principle, 76, 96
Harman, Pamela, 17
Hawks, John, 67, 70
Henrich, J. P., 94
Hill Foundation, 66, 68
"The History of Evolutionary Thought" webpage, 76
Hollinger, Cheryl, 98
Homo naledi fossils, 66f, 67–71
How and Why Wonder Book, 15
Howard Hughes Medical Institute (HHMI), 14, 18, 20, 23, 36, 75,
    76, 89, 98, 100
Howell, Chris, 105
Howell, Patti, 103
"How Evolution Works (and How We Figured It Out)" (PBS Eons,
    video), 83
human body
    evolutionary design of, 33–34
    genetic mutations in, 48, 52
    genetic traits in, 47–48, 54, 93, 96
    skin color, 8, 89–90, 93, 94–95, 99–100 (*see also* race)
    skulls of, 12f, 19f, 43, 44f, 60, 63, 68, 72, 76, 79, 89f, 95f
*Human Errors: A Panorama of Our Glitches, from Pointless Bones to
    Broken Genes* (Lents), 9
Human Evolution Teaching Material Project, 72
"Human Evolution Within the Tree of Life" (HHMI poster), 20
Humes, Edward, 56
Hunter, Lindsay, 65
Hunter, Rick, 65
Hutton, James, 61
Huxley, Thomas, 39, 61, 86

I

iNaturalist (app), 22, 27f
Infidel Guy Show, 49
intelligent design, 2, 58, 60, 107

J

Jablonski, Nina, 26
Jason, Vanessa, 78
Jegede, Olugbemiro J., 57
Jet Propulsion Laboratory, 18
Johanson, Don, 61

Jones, John E., 2
The Journey of Man: A Genetic Odyssey (Wells), 64

K

Keep, Stephanie, 10
Kirshenbaum, Sheril, 107
Kitzmiller v. Dover Area School District (2005), 2
Knuffke, David, 98
Krauss, Lawrence, 77

L

Lamarck, Jean-Baptiste, 61, 83
Langerhans, Brian, 56
The Language of God (Collins), 107
laryngeal nerve (video), 11
Leakey, Louis, 63
Leakey, Louise, 64
Leakey, Richard, 63, 64
Lederman, N. G., 58
Lents, Nathan H., 9
Levine, Joe, 78
Lewontin, Richard, 61, 97
Live Science, 109
logical fallacies, avoiding, 90
Lucy (Johanson), 61
Lyell, Charles, 61, 83

M

macroevolution, 13, 79
Magic of Reality (Dawkins), 19, 87
Making of the Fittest: Natural Selection and Adaptation (video), 17
Malthus, Thomas, 61, 83
Margulis, Lynn, 86, 100
Marr, Kim, 30–31
Martinez, Mary, 4
Matsumoto, Lloyd, 96, 98
Mayr, Ernst, 38
McLean v. Arkansas Board of Education (1982), 2, 45
Mead, John S., 61, 66f, 69f
Meadows, Lee, 107
Meredith, Mark, 43
Metz, Pam, 17
microevolution, 13, 79
Miller, Kenneth, 53, 76, 78, 96, 97, 98
*The Missing Link: An Inquiry Approach for Teaching All Students
    About Evolution* (Meadows), 107
Miss USA 2011—Should Math Be Taught In Schools? (video), 99
Modeling Evolution: The Charlie Shuffle, 78–79
*Monkey Girl: Evolution, Education, Religion, and the Battle for
    America's Soul* (Humes), 56
monkeys, human relationship to, 32, 42, 52, 58, 74, 110. *See also*
    primates, human relationship to
Mooney, Chris, 107

Morgan, Thomas Hunt, 61, 83
Morgan's sphinx hawk moth, 53
Morning, Ann, 97
Morphosource.com, 69
Morris, Hannah, 68
Mowry, David, 73, 78f
Mowry, Ezri and Dipper, 77
Myers, PZ, 49

N
NASA Astrobiology Institute, 18
National Association of Biology Teachers (NABT), 98
National Center for Case Study Teaching in Science, 89, 98
National Center for Science Education (NCSE), 2, 9, 10, 20, 45, 67, 87
National Geographic
    Genographic Project, 64
    Geographic Explorer Classroom program, 69
National Research Council, 109
National Science Foundation, 5, 56
National Science Teachers Association (NSTA), 5, 98, 107
Natterson-Horowitz, Barbara, 18
natural selection. *See also* evolution education
    activity on, guidelines for, 112
    agricultural applications, 5–6, 30
    debate over, 2, 61
    embedding in curriculum, 25
    fittest *vs.* strongest distinction, 52
    introductory explanation of, 10–13, 30, 34–38, 44, 71
    peppered moth simulation, 23, 28, 53
    principles of, 53
    Time Machine (game) on, 54
The Natural Selection (Stated Clearly, video), 49
Natural Selection with the Amoeba Sisters (video), 57
The Nature of Race: How Scientists Think and Teach about Human Difference (Morning), 97
nature of science (NOS)
    definition of, 84
    misconceptions on, 75, 77, 83
    questionnaire on, 58–59, 60
    teaching to understand, 84–87, 88
Nature's Classroom, 73–74
NESCent, 98
Nesse, Randolph, 33
Newton, Isaac, 96
Next Generation Science Standards (NGSS), 18, 25, 26, 34, 84, 89, 98–100, 108

O
Oh Deer! (game), 23
One Long Argument (Mayr), 38
OneZoom website, 9, 54
Only a Theory (Miller), 97
On the Origin of Species (Darwin), 25, 38, 86

On the Origin of Species: The Illustrated Edition (Quammen, illustrator), 25
The Origin of Species: The Beak of the Finch (video), 39, 65
The Origin of Species: The Making of a Theory (video), 75, 76

P
parental opinions, 4, 8–10, 13, 110
PBS Eons team, 83
pedigree charts, 47–48, 52. *See also* cousins, relationships as; evolutionary trees; phylogenetic trees
Peixotto, Becca, 68
Pellock, Brett, 96
peppered moth simulation, 23, 28, 53
Perkinson, Gloria, 62
Perot Museum of Nature and Science, 64, 68
Petto, Andrew J., 1, 111
Pew Research Center, 2, 58
phenotypic traits, 47–48, 54, 96. *See also* skin color/pigmentation
Phillip and Patricia Frost Museum of Science, 6
Phylogenetic Tree of Anole Lizards (activity), 100
phylogenetic trees, 48, 50, 53, 54, 96, 100, 104
Pigliucci, Massimo, 49
Pina, Cara, 96
Pobiner, Briana, 76
poem, on evolution (Duncan), 46
Pony Bead Genetics Lab (activity), 109–110
Pratchett, Terry, 77
primates, human relationship to, 32, 40, 42, 51, 51f, 52, 58, 63, 74, 110
Project WILD, 23
Psychic Network, 49

Q
Quammen, David, 25

R
race, 26, 94–95, 97, 99–100, 102, 110. *See also* skin color/pigmentation
Racism, Not Race: Answers to Frequently Asked Questions (Graves and Goodman), 97
Radford, Benjamin, 50
"A Rainbow of Sepia" (activity), 89–90
randomness, evolution education on, 11–12, 14, 71, 78
Relics of Eden (Fairbanks), 50
religious beliefs
    conflicts due to (*see* conflicts, dealing with)
    on creation (*see* creationism)
    equal time laws, 2, 45, 62
    evolution denial based on, 8–9, 24, 28, 41–42, 50, 62, 74, 79, 105, 110
    evolution education as alternative to, 24, 28, 50, 53, 56, 60, 79–80, 108 (*see also* evolution education)
    evolution education coexistence with, 28, 53, 56–58, 60, 74, 107, 108, 109
    questioning of, 48–49, 103

Remarkable Creatures: Adventures in the Search for the Origin of Species (Carroll), 67

The Richard Dawkins Foundation for Reason & Science, 4–5, 50

The Rise and the Fall of the Dinosaurs (Brusatte), 34, 39

Rising Star Expedition, 65–69

rotation lab (activity), 30–31

Ryan, Mr., 16

S

Sagan, Carl, 17–18

School Board Scenario, 58, 59–60

Schrein, Caitlin, 10

"Science at Multiple Levels" webpage (UC, Berkeley), 76

science education. *See also* nature of science; science teachers

    complexity of, teaching despite, 82, 84–85

    creationism presented as, 2, 45–46, 58, 80

    evidence in (*see* evidence)

    evolution taught in (*see* evolution education)

    misconceptions of, clarifying, 75–76, 90–92, 105–106

    nonscientists' contributions to, 70

    science, defined, 84

    science literacy with, 41, 87, 105–108

    teamwork and collaboration importance to, 70, 83

"Science is the poetry of reality" (Boswell, video), 43

science literacy, 41, 87, 105–108

science teachers. *See also* evolution education; science education

    becoming, paths to, 7, 15–17, 22, 29, 41–43, 48–50, 55–56, 61–63, 73–74, 81–82, 93–94, 103–108

    breadth of topics taught, 3–4, 10, 17, 29

    community of support for, 26, 63, 71, 87–88, 96, 98

    credentials and qualifications of, 7, 10, 32, 50, 62, 74, 82, 99, 103–104

    likability of, 54

    passion and energy of, 53

    professional networking by, 67

    student images of, 7–8

    TIES by/for (*see* Teacher Institute for Evolutionary Science)

sciencetelling approach, 64, 71. *See also* storytelling, power of

Scientists Confront Creationism: Intelligent Design and Beyond (Petto and Godfrey), 1

Scopes, John, trial of, 1, 52, 61, 62

Scott, Eugenie C., 1

Secular Web, 49

The Selfish Gene (Dawkins), 4

Selman, Jeffrey, 53

SETI Institute, 17, 18

sex-linked traits, 47–48

Shook, John, 50

Should Evolution Be Taught in Schools? Worst Top 15 Miss USA Contestant Answers (video), 99

Shubin, Neil, 4, 9, 25, 36, 49

silver foxes, 5f, 6

skin color/pigmentation, 8, 89–90, 93, 94–95, 99–100. *See also* race

The Skull in the Rock (Berger), 64

skulls, 12f, 19f, 43, 44f, 60, 63, 68, 72, 76, 79, 89f, 95f

Smithsonian Institution Human Origins Program, 76, 79

Smithsonian Museum of Natural History, 20

social Darwinism, 94

social justice, 84, 89, 94–95, 99–100, 102

social media

    evaluation of evidence on, 25, 26

    evolutionary discoveries on, 65–66, 69

    networking via, 64–65, 67

    TIES' outreach via, 3, 29, 71

Soto, Patricia, 3–4, 7

speciation, 13, 39, 42, 45, 49, 65, 79, 82, 109

Spot the Misconception (game), 88–89, 92

Stated Clearly, 49

sticky-note histogram project, 28

St. Mark's School of Texas, 63–64

storytelling, power of, 64, 67, 68, 71, 85–87, 91

Student Conservation Association (SCA), 62

students

    classroom as brave space for, 90–91

    creationism education pressure from, 8

    images of science teachers among, 7–8

    questions on evolution, 17, 19

    reflections on evolution education, 53–54

    stress of, from evolution conflicts, 8, 74

    surveying, on evolution, 51–52, 53, 57–60

Subramaniam, Banu, 97

"Survival of the Sneakiest" (story), 100

Symbols, Values, and Norms: Crash Course Sociology #10 (video), 104

Symphony of Science series, 43

T

Taung Child fossil, 65

Teacher Institute for Evolutionary Science (TIES)

    history of, 2–6

    membership of, 1, 29–30, 71, 87, 103

    purpose of, 2–3, 71

    resources from, 3, 5, 10, 12, 18–19, 23, 28, 30–32, 31f, 45, 51, 53, 54, 60, 71, 87–88, 98, 100, 103–104, 108

    Time Machine (game), 14, 54

    workshops by, 3, 5–6, 6f, 12f, 18, 30, 32, 71

teachers. *See* science teachers

Teachers Pay Teachers, 78

teosinte (video), 23

theistic evolution, 58

theory, evolution as, 27–28, 32, 52, 56, 58, 59–60, 76

"This Lab is for the Birds" (activity), 13–14

"Those Pesky Fleas" (bell-ringer), 10–11, 30

3-D printing, 69–70, 72

"thumb" activity, 100

Time Machine (online game), 14, 54

Touchet, Blake, 81, 89f

Tucker, Stephen, 65
Tyson, Neil deGrasse, 41

U

"The Unbroken Thread" (Boswell, video), 43
Understanding Evolution website (UC, Berkeley), 5, 76, 100
Understanding Science website (UC, Berkeley), 76
University of California, Berkeley, 5, 18, 38, 76, 100
University of Edinburgh, 18
University of Hawaii, 18
University of Utah, 45
University of Witwatersrand, 65, 67
Unnatural Selection (van Grouw), 32
Unscientific America—How Scientific Illiteracy Threatens Our Future (Mooney and Kirshenbaum), 107
Upegui, David, 93, 95*f*

V

Vance, M., 95
van Grouw, Katrina, 32
The Variation of Animals and Plants under Domestication (Darwin), 32
Vázquez, Bertha, 1, 5*f*, 7, 12*f*, 18, 29–30, 51, 105, 111
videos
   Bacteria Growing on a Petri Dish, 103
   Blossoms, 100
   The Evolution of Bacteria on a 'Mega-Plate' Petri Dish, 12–13
   Excellent and Simple Explanation of Natural Selection, 12–13
   Facts vs. Theory, 28
   Fact vs. Theory vs. Hypothesis vs. Law Explained, 104
   Fossils, Genes, and Mousetraps, 76–77
   "Great transitions: The Origin of Birds," 36
   "Great transitions: The Origin of Humans," 36
   "Great transitions: The Origin of Tetrapods," 36
   on handwashing, 104
   on historical evolution education (Wallace), 23
   "How Evolution Works (and How We Figured It Out)" (PBS), 83
   on laryngeal nerve, 11
   Making of the Fittest: Natural Selection and Adaptation, 17
   Miss USA 2011—Should Math Be Taught In Schools?, 99
   The Natural Selection (Stated Clearly), 49
   Natural Selection with the Amoeba Sisters, 57
   The Origin of Species: The Beak of the Finch, 39, 65
   The Origin of Species: The Making of a Theory, 75, 76
   Rising Star Expedition interviews, 66
   Rising Star Expedition Twitter Play-by-Play, 65–66
   "Science is the poetry of reality" (Boswell), 43
   Should Evolution Be Taught in Schools? Worst Top 15 Miss USA Contestant Answers, 99
   Symbols, Values, and Norms: Crash Course Sociology #10, 104
   Symphony of Science series, 43
   on teosinte, 23
   "The Unbroken Thread" (Boswell), 43
   "What is the Evidence for Evolution?" (Stated Clearly), 49
   "What is Evolution?" (Stated Clearly), 49
   Why are There Still Chimpanzees?, 40
   Why Evolution is True, Coyne discussing, 50
   "The Wild World of Carnivorous Plants," 31
VISTA (variation, inheritance, selection, time, and adaptation) mnemonic, 38
von Wettberg, Eric, 5–6
Vuletic, Mark, 49

W

Wallace, Alfred Russel, 23, 42, 61, 75, 76, 83, 86
web resources
   "Evolution 101," 76
   "The History of Evolutionary Thought," 76
   iNaturalist (app), 22, 27*f*
   OneZoom, 9, 54
   "Science at Multiple Levels," 76
   Secular Web, 49
   Understanding Evolution, 5, 76, 100
   Understanding Science, 76
Weiner, Jonathan, 34
Wells, Spencer, 64
"What Did *T. rex* Taste Like?" (activity), 38–40
"What is a Theory?" probe, 24
"What is the Evidence for Evolution?" (Stated Clearly, video), 49
"What is Evolution?" (Stated Clearly, video), 49
White, Duncan, 98
White, Tim, 64
Why are There Still Chimpanzees? (video), 40
Why Evolution is True (Coyne), 50
Why We Believe: Evolution and the Human Way of Being (Fuente), 97
Wilberforce, Samuel, 61
"The Wild World of Carnivorous Plants" (video), 31
Williams, George, 33
Wilson, E. O., 2, 94
World Book Encyclopedia, 49
Wright, Sewall, 83

Y

Your Inner Fish: A Journey into the 3.5-Billion-Year History of the Human Body (Shubin), 4, 9, 25, 36

Z

Zoobiquity: The Astonishing Connection Between Human and Animal Health (Natterson-Horowitz), 18

Author Biographies

Nicoline (Nikki) Chambers has taught high school life science (biology, integrated science, and astrobiology) in southern California since 2003. She holds bachelor's and master's degrees in biology from UCLA, and her teaching credential from Cal State Long Beach. She has held teacher ambassador roles for the UC Museum of Paleontology, the Howard Hughes Medical Institute, NASA's Jet Propulsion Lab, and TIES. She also does curriculum development as a "teacher as researcher" at UC Berkeley and the University of Edinburgh. She loves wondering about how life got to be the way it is (on Earth or any other world), and that every answer leads to ten more questions. Nothing makes her happier than seeing the sparks of curiosity and understanding light up a child's eyes.

Amanda Clapp has an MA in anthropology and an MEd in Middle Grades STEM education. She is a National Board Certified teacher and teaches middle school science at The Catamount School, a North Carolina Lab School run by Western Carolina University. The school is operated in cooperation with Jackson County Public Schools, where she has taught in different roles for 16 years. She is a National Geographic Certified Educator and a Kenan Fellow, informing her advocacy for environmental education and equity in North Carolina, and in all schools. As a recipient of the Burroughs Wellcome Fund Career Award for Math and Science Teachers, Amanda is developing a network of partnerships between rural and urban teachers to develop environmental education experiences for their students and strengthen our human understanding through science.

Kenny Coogan earned a BS in animal behavior at the University of Buffalo, NY and an MA in global sustainability. He worked in the education departments of zoological facilities for 10 years before becoming a science teacher. He was awarded the Beginning Science Teacher of the Year Award for the State of Florida through the Florida Association of Science Teachers. He has averaged over $10,000 worth of donations for his classroom per year. In his spare time, he is a regular columnist for print magazines such as *Backyard Poultry*. Kenny shares his one-acre permaculture homestead with cats, chickens, ducks, and a 30-year old Moluccan cockatoo named Buddy. His goal is to live off of the land.

Robert A. Cooper (@bcooper721) recently retired after thirty-six years of teaching science. For the first five years of his career, he taught life science and physical science at the middle school level. For the remaining thirty-one years he taught biology (AP, Honors, and General) at Pennsbury High School, a large high school in the Philadelphia suburbs, earning National Board Certification in 2009. Robert continues to be an advocate for teaching evolution, as he was throughout his teaching career.

Chance Duncan was born and raised in the Arkansas River Valley in Central Arkansas. Chance feels fortunate that he was raised in a rural area and was allowed to explore the fields and woods nearby, getting to know the local ecosystem's flora and fauna. Less lucky for his parents was Chance developing a deep fascination with reptiles, snakes in particular. This fascination led to a tendency to want to share what he found out with his friends and family, so teaching seemed like a natural fit. Chance graduated with his BS in science education from Arkansas Tech University in 2007 and began teaching at a very small, rural school. He moved around to a couple of different districts and began a master's degree program from Montana State University in 2011, completing his MS in science education in 2014. Chance has taught biology in Arkansas for almost a decade and a half and he really can't think of a better career.

Reginald Finley, Sr., is an Atlanta, GA, native. He hosted an online audio program for a decade interviewing experts in diverse scientific fields including Dr. Richard Dawkins. His first experience teaching began as a security trainer, then later as an information technology educator. After earning a bachelor's in human development and a master's in science education, he began tutoring online and volunteered at local science museums. He was eventually afforded an opportunity to work as the director of education and Outreach for Skeleton's Museum in Orlando, Florida. That same year, he was offered a position to work as a biology teacher at Apopka High School. After earning his master's in biology from Clemson, he worked at a private 2nd-12th grade school in Longwood, Florida teaching elementary, middle, and high school students about the wonders of science. He's currently a biology instructor at Valencia College and possesses a PhD in natural science education.

Kathryn (Katie) Green earned her PhD in science education and has spent her life in classrooms as a student, a middle school science teacher, an anthropology instructor, and an educational researcher. She also holds undergraduate and Master's degrees in anthropology and became enthralled with hominid evolution in college. Her mission in life is to support teachers in teaching evolution for understanding and acceptance.

Patti Howell, EdD, has taught high school science for twenty years in Georgia. Prior to teaching, she was a polyurethane chemist. Her commitment to science education does not stop in the classroom. As District 11 Director for Georgia Science Teachers Association, she is an advocate for teacher training and student learning of science in southwest Georgia. She also works as an educator for Albany's Artesian Alliance, which includes Thronateeska Heritage Center, Chehaw Zoo, and Flint Riverquarium, developing curriculum and delivering programs. In her spare time, she loves traveling and hiking with her husband, Seth. Her favorite role, however, is being Granny to Fiona and Lucas.

John S. Mead developed a passion for human origins early in life thanks to books on the topic. John studied at Duke University where he earned his Bachelor and Master of Arts in Teaching (MAT) degrees. Upon graduating in 1990, John found his professional home at the St. Mark's School of Texas where he holds the Eugene McDermott Master Teaching Chair in Science. He has taught most grade levels from 5th to 12th with a focus on the biological sciences and middle school students. In his three decades in the classroom John has worked with a myriad

of scientists connected to evolution as well as the study of human origins. He has traveled to all continents except Antarctica following his love of evolution. In recognition of John's work to expand and improve human origins education he has received awards from both local and national groups. He also works with groups including the National Center for Science Education (NCSE), the American Association of Biological Anthropologists (AABA), and TIES because he believes that science literacy matters. You can find John on Twitter at @Evo_Explorer.

David Mowry was raised and still lives in Bremen, Ohio. (Don't worry, nobody else knows where that is either). He received a BS in wildlife biology from Ohio University in 2004, immediately after which he worked for one of his professors as a field technician on a mark/recapture mammal study. Starting in 2005, David taught outdoor education in Brinkhaven, Ohio for 4 years. He returned to Ohio University to do graduate work in science education and received his teaching certificate. In 2012, he began working as a science instructor at Mid-East Career and Technology Centers in Zanesville, Ohio, where he is still employed. David currently lives in a former funeral home with his wife, two kids, an incredibly stupid dog, and a bunch of cats that won't leave.

Blake Touchet lives in Abbeville, Louisiana with his wife, Chrisanda, and two sons, Luke and Hugo. He has taught middle school, high school, and undergraduate biology and environmental sciences since completing his BS in secondary biology education at the University of Louisiana at Lafayette in 2010. In addition to pursuing advanced degrees in biology (MS from Mississippi State University in 2015) and curriculum leadership (EdD from University of Louisiana at Lafayette in 2021) while teaching, Blake has served as a TIES Teacher Corps member since 2015 and a Teacher Ambassador for the National Center for Science Education since 2017. He has worked on state and district committees for developing curricula, assessments, and mentoring science teachers related to NGSS instructional shifts. His research interests include understanding teacher and administrator knowledge and acceptance of "socially controversial" science topics such as evolution and climate change.

David Upegui is a Latino immigrant who found his way out of poverty through science. He currently serves as a science teacher at his alma mater, Central Falls High School (RI) and as an adjunct professor of education. His personal philosophy and inclusive approach to science education have enabled students to become problem-solvers and innovative thinkers. He has a keen ability to engage students in learning, exploring, and contributing to science. He received the NABT's Evolution Education Award (2014) and the Presidential Award for Excellence in Mathematics and Science Teaching in 2019 (2017 cohort). Upegui started, and runs, the school's Science Olympiad team and has contributed to several publications on science education and appropriate pedagogy. He recently completed his doctoral degree in education at the University of RI, focusing on science education and social justice. Reach Upegui on Twitter (@upeguijara).

Bertha Vázquez has been teaching middle school science in Miami-Dade County Public Schools for 31 years. A seasoned traveler who has visited all seven continents, where she enjoys introducing the world of nature and science to young, eager minds. An educator with National Board Certification, she is the recipient of several national and local honors, including the 2014 Samsung's $150,000 Solve For Tomorrow Contest and the 2009 Richard C. Bartlett Award for excellence in environmental education. She was Miami-Dade County Public School's Science Teacher of the Year in 1997, 2008, and 2017. Thanks to her work with TIES, she was also the 2017 winner of the National Association of Biology Teachers Evolution Education Award. Bertha has been the director of the Teacher Institute for Evolutionary Science, a project of the Center for Inquiry, since 2015. In her spare time, Bertha is an avid tennis player and bashful ukulele player. She lives in Miami, Florida with her husband, son, and two dogs.

Acknowledgments

Truth be told, the success of The Teacher Institute for Evolutionary Science has taken me by surprise. Not even in my wildest dreams did I expect this project to grow the way it has. I thank Richard Dawkins for his vision and support, he is ever willing to help TIES any way he can. The staff at The Center for Inquiry (CFI) has been incredible. Any request I make about the webpage or TIES resources are taken care of so quickly it makes my head spin. I am grateful to CFI's President and CEO, Robyn Blumner, who has placed tremendous confidence in me. She has allowed me to guide this project from the perspective of a classroom teacher. And, of course, I am thankful for all of my fellow TIES educators and my fellow chapter authors, especially Kenny Coogan, who is a big part of our success. One of the exceptional educators who joined us very early on is John Mead, Eugene McDermott Master Teacher of Science at St. Mark's School of Texas in Dallas. John has great interest in the study of human evolution and how its study can be used to inspire students. It was John's idea to write a book about TIES and evolution education. We owe this book project to him. Finally, I must thank Dr. Lawrence Bonchek, the generous donor and TIES supporter who made the publishing of this book possible.

Acknowledgments